Thermoelectric Micro/Nano Generators 2

SCIENCES

Energy, Field Directors – Alain Dollet and Pascal Brault

Energy Recovery, Subject Head – Gustavo Ardila

Thermoelectric Micro/Nano Generators 2

Challenges and Prospects

Coordinated by
Hiroyuki Akinaga
Atsuko Kosuga
Takao Mori
Gustavo Ardila

iSTE

WILEY

First published 2023 in Great Britain and the United States by ISTE Ltd and John Wiley & Sons, Inc.

ISTE Ltd
27-37 St George's Road
London SW19 4EU
UK

www.iste.co.uk

John Wiley & Sons, Inc.
111 River Street
Hoboken, NJ 07030
USA

www.wiley.com

Library of Congress Control Number: 2023942071

British Library Cataloguing-in-Publication Data
A CIP record for this book is available from the British Library
ISBN 978-1-78945-145-0

ERC code:
PE8 Products and Processes Engineering
PE8_6 Energy processes engineering

Contents

Preface ix
Hiroyuki AKINAGA, Atsuko KOSUGA and Takao MORI

Introduction xiii
Hiroyuki AKINAGA, Atsuko KOSUGA and Takao MORI

Part 1. Material Challenges and Novel Effects 1

Chapter 1. Reliability and Durability of Thermoelectric Materials and Devices: Present Status and Strategies for Improvement 3
Congcong XU, Hongjing SHANG, Zhongxin LIANG, Fazhu DING and Zhifeng REN

1.1. Introduction 3
1.2. Thermoelectric material stability 5
1.3. $Mg_3(Sb, Bi)_2$ 5
1.4. Zn_4Sb_3 7
1.5. Skutterudites 8
1.6. $Cu_{2-x}X$ (X = S, Se, Te) 9
1.7. GeTe 11
1.8. Outlook on thermoelectric materials stability 12
1.9. Thermoelectric device design analysis 13
1.9.1. Thermal stress analysis 13
1.9.2. Interface analysis, design and fabrication 21
1.10. Advanced thermoelectric module case studies 33
1.10.1. Bi_2Te_3 33
1.10.2. $Mg_3(Sb, Bi)_2$ 35
1.10.3. GeTe 37
1.10.4. Skutterudites 39
1.11. Summary and outlook 40
1.12. References 41

Chapter 2. Effect of Microstructure in Understanding the Electronic Properties of Complex Materials 53
Chenguang FU, Chaoliang HU, Qi ZHANG, Airan LI and Tiejun ZHU

2.1. Introduction 53
2.2. Basic principles of electronic transport parameters 54
2.2.1. Solid solutions 59
2.2.2. Intrinsic defects 60
2.2.3. Grain boundary 62
2.2.4. Texture 65
2.3. Summary 67
2.4. References 67

Chapter 3. Thermoelectric Nanowires 73
Olga CABALLERO-CALERO and Marisol MARTÍN-GONZÁLEZ

3.1. Introduction 73
3.2. Nanowires: a way to enhance thermoelectric efficiency 74
3.3. Fabrication of thermoelectric nanowires 77
3.4. Measurement of thermoelectric properties in nanowires 79
3.5. Nanowire-based thermoelectric devices 86
3.6. Interconnected 3D nanowire networks 87
3.7. Summary and outlook 89
3.8. References 89

Chapter 4. Impact of Chemical Doping or Magnetism in Model Thermoelectric Sulfides 99
Sylvie HÉBERT, Ramzy DACU and Antoine MAIGNAN

4.1. Introduction 100
4.2. TiS_2: intercalation chemistry to combine power factor optimization and lattice thermal conductivity degradation 101
4.3. Magnetism and thermoelectricity in sulfides 104
4.4. Conclusion 110
4.5. References 110

Chapter 5. Thermoelectric Generation Using the Anomalous Nernst Effect 117
Akito SAKAI and Satoru NAKATSUJI

5.1. Thermoelectric conversion – Seebeck effect and anomalous Nernst effect (ANE) 117
5.2. Physics of topological magnets 120
5.2.1. Transverse electrical and thermal conductivity driven by Berry curvature 120
5.2.2. Magnetic Weyl semimetals, Weyl magnets 121

5.2.3. Type-II Weyl semimetals . . . 122
5.2.4. Nodal line magnets . . . 123
5.3. Experimental realization of the giant anomalous Nernst effect . . . 124
5.3.1. Weyl antiferromagnets Mn_3X (X = Sn, Ge) . . . 124
5.3.2. Weyl ferromagnet Co_2MnGa . . . 124
5.3.3. Nodal-web ferromagnets Fe_3X (X = Ga, Al) . . . 125
5.4. Summary and prospects . . . 127
5.5. Acknowledgment . . . 127
5.6. References . . . 127

Chapter 6. A Comprehensive Review of Phonon Engineering . . . 131
Bin XU, Harsh CHANDRA, and Junichiro SHIOMI

6.1. Introduction . . . 131
6.1.1. Thermal conductivity . . . 133
6.1.2. Phonons in thermal transport. . . . 133
6.2. Methodology of phonon engineering . . . 142
6.2.1. Computational method for thermal conduction and phonon properties . . . 142
6.2.2. Experimental method for nano-/micro-scale heat conduction characterization. . . . 144
6.2.3. Direct measurement of phonon properties through phonon scattering . . . 149
6.2.4. Phonon engineering for low thermal conductivity . . . 152
6.2.5. Intrinsic low thermal conductivity in complex lattice structure . . . 153
6.2.6. Low thermal conductivity by nanostructures . . . 155
6.2.7. Coherent phonon engineering in superlattice . . . 158
6.3. Summary and future prospects . . . 162
6.4. References . . . 164

Part 2. Toward Device Applications . . . 171

Chapter 7. The Current State of Thermoelectric Technologies and Applications with Prospects . . . 173
Slavko BERNIK

7.1. Introduction . . . 173
7.2. Thermoelectric materials . . . 180
7.3. Thermoelectric devices – structure, materials, fabrication technology . . . 189
7.4. Summary . . . 198
7.5. References . . . 199

Chapter 8. Processing of Thermoelectric Transition Metal Silicides Towards Module Development 213
Sylvain LE TONQUESSE, Mathieu PASTUREL, Franck GASCOIN and David BERTHEBAUD

8.1. Introduction 213
8.2. Recent progress on the process of thermoelectric transition metals silicide . . 214
8.2.1. Synthesis of mesostructured silicides through magnesiothermic reduction 214
8.2.2. Synthesis of higher manganese silicide through wet ball milling. 218
8.2.3. Issues of MnSi striations and thermal stability on thermoelectric performance of doped higher manganese silicide 219
8.2.4. Upscaling processes, the examples of additive manufacturing and RGS process 223
8.3. Towards contacts and device developments. 225
8.4. References 226

Chapter 9. Application of the Thermoelectrics; Past, Present and Future 229
Hirokuni HACHIUMA

9.1. Introduction 229
9.2. Thermoelectric module 230
9.3. TEC application for refrigerator and cooler 231
9.4. TEC for electronic components. 234
9.4.1. TEC for optical communication 234
9.4.2. Multi-stage TEC for optical sensors. 236
9.5. TEC for semiconductor manufacturing 239
9.6. TEG application 241
9.6.1. TEG for energy harvesting (EH). 242
9.6.2. TEG for stand-alone power source 244
9.6.3. TEG for waste heat recovery. 245
9.7. Conclusion 247
9.8. References 247

List of Authors 249

Index. 253

Summary of Volume 1 255

Preface

Hiroyuki AKINAGA[1], Atsuko KOSUGA[2] and Takao MORI[3,4]
[1] *National Institute of Advanced Industrial Science and Technology (AIST), Device Technology Research Institute, Japan*
[2] *Osaka Metropolitan University, Department of Physical Science, Graduate School of Science, Japan*
[3] *National Institute for Materials Science (NIMS), WPI-MANA, Tsukuba, Japan*
[4] *Graduate School of Pure and Applied Sciences, University of Tsukuba, Japan*

Thermoelectric materials can be utilized in solid state devices (i.e. thermoelectric generators) to convert heat to electricity. They therefore are of high potential interest, in usages for conversion of waste heat to save energy, and as convenient and maintenance-free sources to power the huge number of Internet of Things (IoT) sensors. With this background in mind, the two volume edition of *Thermoelectric Micro/Nano Generators* has been brought together, to serve as an important, comprehensive set of books encompassing the fundamental principles, state of the art advancements, and outlooks for this topic, as also noted in the Preface of *Thermoelectric Micro/Nano Generators 1*. Volume 1 mainly dealt with the fundamental physics, materials and measurements of thermoelectrics.

In this volume, Volume 2, we will mainly deal with the challenges and prospects, especially in relation to thermoelectric devices and applications, while also introducing some non-conventional attempts. As presented in

Thermoelectric Micro/Nano Generators 2,
coordinated by Hiroyuki AKINAGA, Atsuko KOSUGA,
Takao MORI and Gustavo ARDILA.

Volume 1, the performance of thermoelectric materials has greatly advanced in recent years, as has the development of sustainable, abundant elemental materials. There are still various aspects that must be addressed in order to bring thermoelectric materials to viable energy conversion applications, such as various applicative considerations related to devices, e.g. durability, processing, design, and identification of suitable applications. These are covered in this volume. After a brief introduction on the landscape of thermoelectric generators in energy harvesting technologies by several of the editors, a comprehensive chapter titled "Reliability and Durability of Thermoelectric Materials and Devices: Present Status and Strategies for Improvement" (Chapter 1) is given by Congcong Xu, Hongjing Shang, Zhongxin Liang, Fazhu Ding, and Zhifeng Ren. They present in detail several key issues to bridge the gap between fundamental thermoelectric research and the succesful implementation to thermoelectric applications. Firstly, they overview the thermal stability of several prominent, sustainable, high performance, new generation thermoelectric materials, i.e. $Mg_3(Sb.Bi)_2$-type, Zn_4Sb_3, skutterudites, copper chalcogenides, and also GeTe, which can be the lead-free replacement to PbTe. They analyze any thermal or material stability issues and give routes for improvement of stability. Secondly, as the main part, they give a detailed analysis of thermal device design, since it is critically important to construct devices that are thermally and mechanically reliable. They consider thermal stress, such as from thermal cycling and shock, interface issues, such as contact, barrier layers and joining technologies, etc. and importantly, they give comprehensive guidelines for particular designs and strategies to address these issues. Finally, they give several illustrative detailed thermoelectric module case studies.

Chapter 2, written by Chenguang Fu, Chaoliang Hu, Qi Zhang, Airan Li, and Tiejun Zhu, deals with "Effect of Microstructure in Understanding the Electronic Properties of Complex Materials". As also reviewed in Volume 1, microstructuring has been demonstrated to be a very powerful method to selectively scatter phonons and enhance the figure of merit ZT. In this chapter, the authors focus on the detailed effects of such microstructuring on the electronic properties that are less well established. They demonstrate how to gain insight into this from experimental data of the Seebeck coefficient, electrical conductivity, mobility, and discuss strategies on how to minimize the detrimental effects on the electronic transport from point defects, grain boundaries texturing, etc., which are important for effective phonon scattering. The next three chapters cover developments of some novel systems/mechanisms. Olga Caballero-Calero and Marisol Martín-González review the developments in thermoelectric nanowires

(Chapter 3). Nanowires are of interest for both thermoelectric and thermal conductivity aspects. The famous quantum confinement effect proposed by Hicks and Dresselhaus particularly targets nanowires, since it was proposed that in low dimensional nanostructures, the sharp features of the energy derivative of density of states could be utilized to enhance the Seebeck coefficient. Furthermore, the nanostructure could lower the thermal conductivity. Here the authors comprehensively present the fabrication methods, measurements and thermoelectric properties of nanowires regarding a wide variety of materials. They also cover nanowire-based devices and the outlook for these morphology materials and devices. Sylvie Hébert, Ramzy Daou, and Antoine Maignan cover the impact of chemical doping or magnetism in model thermoelectric sulfides (Chapter 4). They particularly focus on the intercalation stategy in the layered sulfide TiS_2, and as a broad principle, the role that magnetism can play to enhance the Seebeck coefficient. Sulfides have followed oxides in being an interesting playground where magnetism can play a direct role in enhancing the entropy part of the Seebeck coefficient, for example. Such a mechanism is at least partly indicated for the $CuCrTiS_4$ spinel, and doped CoS_2 exhibits an interesting spin polarization dependence. Related to the magnetism and thermoelectricity theme, in Chapter 5, Akito Sakai and Satoru Nakatsuji review thermoelectric generation using the anomalous Nernst effect (ANE). While the anomalous Nernst effect is still much smaller than the Seebeck effect, recent advancements have shown a large increase in the effect from much smaller levels, and mechanisms such as dependence on the Berry curvature, Weyl points, etc. are becoming clearer. This chapter reviews such fundamental physics, and covers various topological magnets such as magnetic Weyl semimetal, Weyl magnet examples showing promising ANE features. In the final chapter of this section, Bin Xu, Harsh Chandra, and Junichiro Shiomi have written a useful counterpart to the previous chapter on the microstructure effects on the electrical transport with a comprehensive review of phonon engineering (Chapter 6). They review fundamentals in phonon transport, various methodology in measurements and computation, and phonon engineering such as via superlattices. One particular emphasis of the chapter is on discussing the breakdown and analysis of the anharmonic properties of phonons leading to low thermal conductivity.

The final three chapters deal with thermoelectric devices and applications. The first by Slavko Bernik kicks off this section by giving an introductory overview of the current state of thermoelectric technologies and applications with prospects (Chapter 7). Starting with the background of thermoelectric applications, the author gives a broad overview on the various materials classes and discussions on considerations for applicability, and also goes into detail regarding the basic structure and various requirements for devices, including fill factor, integration, packaging, etc., and also considerations for the different power ranges for applications. Sylvain Le Tonquesse, Mathieu Pasturel, Franck Gascoin, and David Berthebaud write a detailed and practical treatise entitled "Processing of Thermoelectric Transition Metal Silicides Towards Module Development" (Chapter 8). Transition metal silicides have attracted attention because their respectable, while not ground-breaking, thermoelectric performance is augmented largely by their being very abundant, inexpensive, materials with generally good mechanical properties. The authors particularly focus on synthesis and processing methods with a view to being compatible to industry, and present stage of development of contacts and devices. In the final chapter, Hirokuni Hachiuma gives a comprehensive and detailed review of the application of thermoelectrics in the past, the present, and the future (Chapter 9). He first covers the early development of the thermoelectric refrigerator and cooler, which have grown to various applications for thermoelectric cooling and present large growth of the market. The author then covers the thermoelectric power generation applications that are being developed, for categories such as energy harvesting, stand-alone power sources, and waste heat recovery.

This second volume on thermoelectric micro/nano generators comprehensively covers the fundamentals and state-of-the-art advances in thermoelectric device and material applicative requirements in particular, together with several novel directions. This volume should be especially useful to newcomers to the field, who have an interest in possible applications for example coming from industry, or have interest in the novel topics, as well as experts desiring guidelines for the advanced development of various materials' electrical and thermal transport, and mastering various requirements and strategies for devices and applications. Coupled with the first volume, this two volume set covers the fundamentals to practical issues and the state-of-the-art thermoelectrics field, which should be of value to be studied and utilized by many readers.

July 2023

Introduction

Landscape of Thermoelectric Generators in Energy Harvesting Technologies

Hiroyuki AKINAGA[1], Atsuko KOSUGA[2] and Takao MORI[3,4]

[1] *National Institute of Advanced Industrial Science and Technology (AIST), Device Technology Research Institute, Japan*

[2] *Osaka Metropolitan University, Department of Physical Science, Graduate School of Science, Japan*

[3] *National Institute for Materials Science (NIMS), WPI-MANA, Tsukuba, Japan*

[4] *Graduate School of Pure and Applied Sciences, University of Tsukuba, Japan*

What is the essential infrastructure of our society? It is an Internet of Things (IoT) network where we can communicate and receive information. Nowadays it is possible to connect to every corner of the world with a smartphone. By sending and receiving information day and night across national borders, we can see what is happening around the world. Of course, in order to use information terminals such as smartphones, we need electric power. In everyday life, electric power can be obtained from outlets. Where

Thermoelectric Micro/Nano Generators 2,
coordinated by Hiroyuki AKINAGA, Atsuko KOSUGA,
Takao MORI and Gustavo ARDILA.

there is no outlet, for example, when traveling in the woods, you will probably carry a charger with you. However, it is not always possible to get enough electricity. If there is a natural disaster such as a typhoon or an earthquake, or if society becomes unstable for whatever reason, the power supply will stop for a long time. We must be able to run smartphones and generate enough power to communicate when needed. A thermoelectric generator is a technology that uses the heat around us to generate electricity.

The dissipating heat can be harvested and converted into electric energy by thermoelectric energy harvesting (Snyder and Toberer 2008; Champier 2017; O'Dwyer et al. 2017; Siddique et al. 2017; Akinaga et al. 2018; Haras and Skotnicki 2018; Kishore and Priya 2018; Mori and Priya 2018; Petsagkourakis et al. 2018; Yan et al. 2018; Akinaga 2020; Hewawasam et al. 2020; Jaziri et al. 2020; Nandihalli et al. 2020; Zoui et al. 2020; Tohidi et al. 2022). This effect is called the Seebeck effect, and when a temperature difference ΔT is given at two ends of a material, a voltage ΔV is generated as proportional to ΔT. The proportionality coefficient of this phenomenon S (= ΔV / ΔT) is called the Seebeck coefficient. The Seebeck effect occurs because the charge carriers diffuse when a temperature difference is applied to the material, which results in the electrical voltage, ΔV, to be generated. Thermoelectric materials are frequently evaluated by the dimensionless figure of merit $ZT = (S^2\sigma/\kappa)T$, where σ is the electrical conductivity, κ is the thermal conductivity, and T is the absolute temperature. $S^2\sigma$ [W/m/K^2] in the numerator is called the power factor (PF) and indicates how much power can be obtained with a temperature difference of 1°C per unit length of the device under test.

The thermoelectric conversion efficiency is given by $\eta = \eta_{\mathrm{Carnot}}[(1 + ZT_{\mathrm{m}})^{1/2} - 1] / [(1 + ZT_{\mathrm{m}})^{1/2} + T_{\mathrm{c}}/T_{\mathrm{h}}]$, where the ideal Carnot efficiency $\eta_{\mathrm{Carnot}} = (T_{\mathrm{h}} - T_{\mathrm{c}})/T_{\mathrm{h}}$, T_{h} and T_{c} are the hot side and cold side temperatures, respectively, and T_{m} is the average temperature.

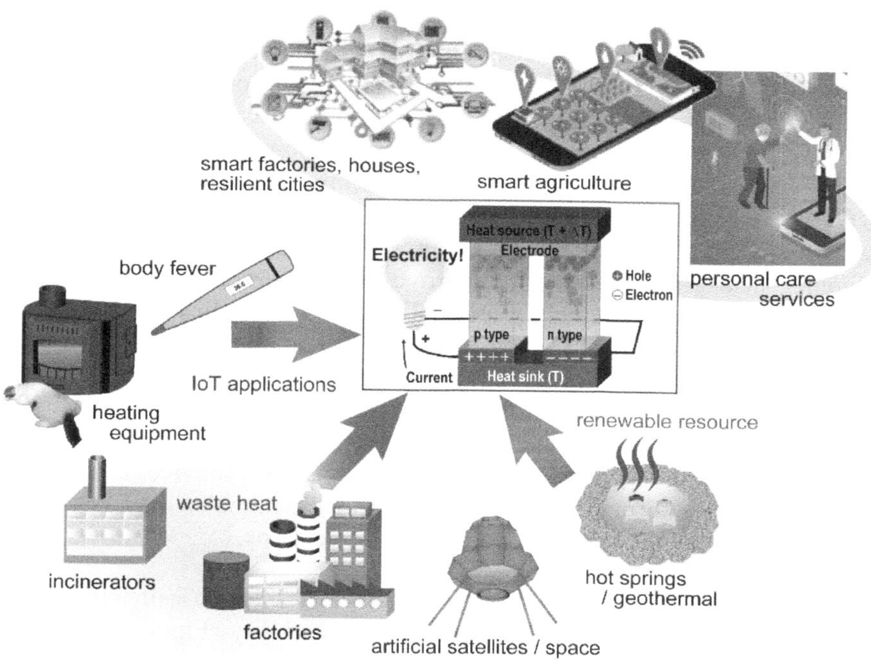

Figure I.1. *Landscape of Thermoelectric Generators*

The long-standing issue in achieving high conversion efficiency in thermoelectric materials is how ZT has competing requirements in the physical properties of which it is comprised. It could be thought that the properties of thermoelectric materials should be metallic, because of σ in the numerator, and from an application standpoint, in order to decrease the contact resistance of the electrodes. However, ordinary metals cannot achieve a high thermoelectric conversion efficiency because the number of charge carriers does not greatly change with the temperature, and thereby Seebeck coefficients are typically small. Furthermore, S, σ and κ in the equation for ZT are functions of carrier concentration and are difficult to control independently. Therefore, in order to achieve high thermoelectric performance, it is necessary to find a way to overcome these material paradoxes (Qian et al. 2021; Hendricks et al. 2022). For example, researchers have attempted to improve the effective thermoelectric properties of materials by utilizing nanotechnology and multi-disciplinary concepts. To give a couple non-conventional examples, the control of thermal conduction in phononic crystals with a periodic nanostructure is an interesting approach (Nomura et al. 2018). The utilization of magnetic spin degrees of freedom is also a promising method (Koshibae et al. 2000; Uchida et al. 2008; Hendricks et al. 2022). Although various thermoelectric materials have been reported, a considerable number of them show a relatively large decrease in *ZT* near room temperature (RT). This is why Bi_2Te_3 has typically been almost the only choice of thermoelectric material near RT. It is also difficult to obtain a sufficiently large temperature difference near RT. Research to improve the properties of thermoelectric materials near RT is extremely important for the IoT application, and recent advances have been made (Wang et al. 2020; Liu et al. 2021, 2022; Oku et al. 2021). To increase the use of thermoelectric generators in society, it is also necessary to establish techniques for evaluating the various thermoelectric properties of materials and devices with a sufficiently high accuracy as well as reproducibility. The development of device, module and upscaling manufacturing processes developments, are also key issues for thermoelectric generators to achieve viable applications.

If thermoelectric generator can replace batteries for IoT, it will have a huge social impact from the viewpoint of sustainability (Figure I.1). It will also enable us to deliver affordable, reliable and sustainable electricity to people around the world. Considering the amount of electricity, thermoelectric generators will contribute not only to the IoT field but rather to industries and transportation. It is well known that about 70% of the global primary energy is lost. Most of the energy is lost as waste heat and dumped into the environment (Forman et al. 2016). Although the temperature range of waste heat can be over 300 degrees, for a variety of applications, about 60% of the waste heat streams arise at a temperature below 100°C. In addition, global energy consumption is expected to increase by 40% by 2040 compared to 2015, according to International Energy Outlook (U.S. Energy Information Administration 2017). This outlook shows that the industrial sector continues to account for the largest share of energy consumption.

The worldwide increasing demand in energy has increased social unrest and political tensions. Environmental impact of global climate change from the use of fossil fuels is among the most well-known challenges that humanity has ever faced. As a measure to solve these problems, diversified renewable energies are being promoted. There is a great tendency towards renewable energy sources, but unlike fossil fuel energy sources, renewable energy sources are unreliable and intermittent, requiring technological development to utilize them in a complementary way. Thus, our society needs stable and sustainable electricity. As a means to solve the abovementioned energy problems, the thermoelectric generator, which uses a large amount of waste heat that is constantly being generated, will become indispensable to our society as a stable and clean source of electricity.

References

Akinaga, H. (2020). Recent advances and future prospects in energy harvesting technologies. *Jpn. J. Appl. Phys.*, 59, 110201.

Akinaga, H., Fujita, H., Mizuguchi, M., Mori, T. (2018). Focus on advanced materials for energy harvesting: Prospects and approaches of energy harvesting technologies. *Sci. Technol. Adv. Mater.*, 19, 543–544.

Champier, D. (2017). Thermoelectric generators: A review of applications. *Energy Convers. Manag.*, 140, 167–181.

Forman, C., Muritala, I.K., Pardemann, R., Meyer, B. (2016). Estimating the global waste heat potential. *Renew. Sust. Energ. Rev.*, 57, 1568–1579.

Haras, M. and Skotnicki, T. (2018). Thermoelectricity for IoT – A review. *Nano Energy*, 54, 461–476.

Hendricks, T., Caillat, T., Mori, T. (2022). Keynote review of latest advances in thermoelectric generation materials, devices, and technologies 2022. *Energies*, 15, 7307.

Hewawasam, L.S., Jayasena, A.S., Afnan, M.M.M., Ranasinghe, R.A.C.P., Wijewardane, M.A. (2020). Waste heat recovery from thermo-electric generators (TEGs). *Energy Rep.*, 6, 474–479.

Jaziri, N., Boughamoura, A., Müller, J., Mezghani, B., Tounsi, F., Ismail, M. (2020). A comprehensive review of thermoelectric generators: Technologies and common applications. *Energy Rep.*, 6, 264–287.

Kishore, R.A. and Priya, S. (2018). A review on low-grade thermal energy harvesting: Materials, methods and devices. *Materials (Basel)*, 11, 1433.

Koshibae, W., Tsutsui, K., Maekawa, S. (2000). Thermopower in cobalt oxides. *Phys. Rev. B*, 62, 6869–6872.

Liu, Z., Sato, N., Gao, W., Yubuta, K., Kawamoto, N., Mitome, M., Kurashima, K., Owada, Y., Nagase, K., Lee, C.-H. et al. (2021). Demonstration of ultrahigh thermoelectric efficiency of ~ 7.3% in Mg_3Sb_2/MgAgSb module for low-temperature energy harvesting. *Joule*, 5, 1196–1208.

Liu, Z., Gao, W., Oshima, H., Nagase, K., Lee, C.H., Mori, T. (2022). Maximizing the performance of n-type Mg_3Bi_2 based materials for room-power generation and thermoelectric cooling. *Nat. Commun.*, 13, 1120.

Mori, T. and Priya, S. (2018). Materials for energy harvesting: At the forefront of a new wave. *MRS Bull.*, 43, 176–180.

Nandihalli, N., Liu, C.-J., Mori, T. (2020). Polymer based thermoelectric nanocomposite materials and devices: Fabrication and characteristics. *Nano Energy*, 78, 105186.

Nomura, M., Shiomi, J., Shiga, T., Anufriev, R. (2018). Thermal phonon engineering by tailored nanostructures. *Jpn. J. Appl. Phys.*, 57, 080101.

O'Dwyer, C., Chen, R., He, J.-H., Lee, J., Razeeb, K.M. (2017). Scientific and technical challenges in thermal transport and thermoelectric materials and devices. *ECS J. Solid State Sci. Technol.*, 6, N3058–N3064.

Oku, T., Funashima, H., Kawaguchi, S., Kubota, Y., Kosuga, A. (2021). Superior room-temperature power factor in GeTe systems via multiple valence band convergence to a narrow energy range. *Mater. Today Phys.*, 20, 100484.

Petsagkourakis, I., Tybrandt, K., Crispin, X., Ohkubo, I., Satoh, N., Mori, T. (2018). Thermoelectric materials and applications for energy harvesting power generation. *Sci. Technol. Adv. Mater.*, 19, 836–862.

Qian, X., Zhou, J., Chen, G. (2021). Phonon-engineered extreme thermal conductivity materials. *Nat. Mater.*, 20, 1188–1202.

Siddique, A.R.M., Mahmud, S., Heyst, B.V. (2017). A review of the state of the science on wearable thermoelectric power generators (TEGs) and their existing challenges. *Renew. Sust. Energ. Rev.*, 73, 730–744.

Snyder, G.J. and Toberer, E.S. (2008). Complex thermoelectric materials. *Nat. Mater.*, 105–114.

Tohidi, F., Ghazanfari Holagh, S., Chitsaz, A. (2022). Thermoelectric generators: A comprehensive review of characteristics and applications. *Appl. Therm. Eng.*, 201, 117793.

Uchida, K., Takahashi, S., Harii, K., Ieda, J., Koshibae, W., Ando, K., Maekawa, S., Saitoh, E. (2008). Observation of the spin Seebeck effect. *Nature*, 455, 778–81.

U.S. Energy Information Administration (2017). International Energy Outlook 2017 [Online]. Available at: https://www.eia.gov/outlooks/ieo/pdf/0484(2017).pdf [Accessed 6 June 2022].

Wang, L., Li, J., Zhang, C., Ding, T., Xie, Y., Li, Y., Liu, F., Ao, W., Zhang, C. (2020). Discovery of low-temperature GeTe-based thermoelectric alloys with high performance competing with Bi_2Te_3. *J. Mater. Chem. A*, 8, 1660–1667.

Yan, J., Liao, X., Yan, D., Chen, Y. (2018). Review of micro thermoelectric generator. *J. Microelectromech. Syst.*, 27, 1–18.

Zoui, M.A., Bentouba, S., Stocholm, J.G., Bourouis, M. (2020). A review on thermoelectric generators: Progress and applications. *Energies*, 13, 3606.

PART 1

Material Challenges and Novel Effects

1

Reliability and Durability of Thermoelectric Materials and Devices: Present Status and Strategies for Improvement

Congcong XU[1,*], Hongjing SHANG[1,2,*], Zhongxin LIANG[1,*], Fazhu DING[2], and Zhifeng REN[1]

[1] *Department of Physics and Texas Center for Superconductivity (TcSUH), University of Houston, USA*

[2] *Key Laboratory of Applied Superconductivity, Institute of Electrical Engineering, Chinese Academy of Sciences, Beijing, China*

**These authors contributed equally to this chapter.*

1.1. Introduction

The performance of a thermoelectric material is usually evaluated by its operating temperature (T) and three other interrelated parameters: electrical conductivity (σ), Seebeck coefficient (S) and thermal conductivity (k). These parameters together determine the thermoelectric dimensionless figure of merit (zT), $zT = \frac{S^2\sigma}{\kappa}T$. The thermoelectric conversion efficiency is

Thermoelectric Micro/Nano Generators 2,
coordinated by Hiroyuki AKINAGA, Atsuko KOSUGA,
Takao MORI and Gustavo ARDILA.

dependent on the zT. Theoretically, higher zT values can guarantee higher thermoelectric conversion efficiency, so there has been a great effort to improve zT over the past several decades, and the zT of some materials has been reported exceeding 2. Nevertheless, the development of thermoelectric devices has been somewhat lagging in comparison to the performance optimization of materials.

Additional scientific and engineering challenges are faced during thermoelectric device development, including long-term material stability, material parameter matching, contact layer selection, device structure and thermoelectric system design, etc. Among these challenges, stability is a prerequisite for the actual application of materials. Since thermoelectric devices are operated under large temperature differences, the environment faced by thermoelectric materials, especially in thermoelectric power generators, is relatively harsh. Repeated thermal cycles, thermal shocks and high temperatures will not only reduce the thermomechanical properties of a material but may also change its composition, which ultimately leads to deterioration in the thermoelectric performance and even complete device failure. In addition, the potential thermal stress caused by repeated thermal expansion and contraction imposes more stringent requirements on the stability of the device interface and the thermomechanical matching of the p- and n-type materials. Unfortunately, in comparison to the effort to improve zT, less attention has been paid to the stability issue for thermoelectric materials and devices, limiting their practical application.

Here, we will highlight this issue and bridge the gap between thermoelectric research and actual applications by discussing the durability and reliability challenges of both thermoelectric materials and devices. We first address the thermal stability of several thermoelectric materials and strategies for thermal stability improvement. Secondly, we analyze in detail thermoelectric device design, including thermal stress and interface issues. Finally, we summarize recent progress in the design and fabrication of reliable and durable thermoelectric modules and provide an outlook for future research needs.

1.2. Thermoelectric material stability

For actual applications, a material's stability is at least as critical as its thermoelectric properties, if not more important. Different kinds of instability, including component decomposition, elemental evaporation and phase transition, can result in the degradation of thermoelectric material performance, further leading to the decreased energy-conversion efficiency of thermoelectric modules. In the following section, typical instability issues of several materials with decent thermoelectric performances are discussed in detail. Strategies that have been used to improve the thermal stability of these materials are also summarized and discussed.

1.3. $Mg_3(Sb, Bi)_2$

$Mg_3(Sb, Bi)_2$ exhibits a $CaAl_2Si_2$-type crystal structure consisting of an octahedrally coordinated Mg^{2+} cation layer and a tetrahedrally coordinated anion structure $[(Mg_2Sb_2)^{2-}]$ (Shuai et al. 2017a; Song et al. 2019; Mao et al. 2018). It had long been considered a p-type semiconductor with undistinguished performance, but Te-doped n-type $Mg_3(Sb, Bi)_2$-based alloys were recently reported to show high thermoelectric performance, demonstrating good potential for both power generation and cooling applications (Tamaki et al. 2016; Zhang et al. 2017). Subsequently, various strategies, including tuning the carrier scattering mechanism (Mao et al. 2017a; Shuai et al. 2017b), defect engineering (Mao et al. 2017b; Li et al. 2019, 2020a) and doping (Shi et al. 2019b, 2019c; Shu et al. 2019), have been successfully applied in Mg_3Sb_2-based alloys, achieving a state-of-the-art average zT value up to 1.1 in the range of 300~500 K (Imasato et al. 2019a; Shi et al. 2019a; Shang et al. 2020a; Wood et al. 2019).

Among all defects in Mg_3Sb_2-based materials, Mg vacancies have the lowest defect formation energy around the Fermi level, thus resulting in the intrinsically high concentration of Mg vacancies causing p-type conduction of these materials (Zhang et al. 2019). The addition of extra Mg into the matrix has been reported to be successful in reducing the Mg-vacancy concentration and achieving n-type conduction (Shuai et al. 2018; Kuo et al. 2019). At the operating temperature of 773 K, the corresponding vapor pressures of elemental Mg, Sb and Bi are 7,390 Pa, 139 Pa and 9.64 Pa, respectively. Such a great difference in vapor pressure will result in significant Mg loss at high temperatures. Jørgensen et al. (2018) studied the thermal stability of $Mg_3Sb_{1.475}Bi_{0.475}Te_{0.05}$ via high-temperature synchrotron powder X-ray diffraction and found that a Bi impurity phase forms in thermal cycles from 300 K to 725 K, suggesting the compound's high-temperature structural instability. Shang et al. (2020b) reported that $Mg_3Sb_{2-x}Bi_x$ alloys ($Mg_{3.2}Sb_{1.5}Bi_{0.49}Te_{0.01}$, $Mg_{3.2}SbBi_{0.99}Te_{0.01}$ and $Mg_{3.2}Sb_{0.49}Bi_{1.5}Te_{0.01}$) showed performance degradation at 673 K and 773 K via long-term in situ measurements of their thermoelectric properties, although they demonstrated the stable electrical properties of these compounds at 573 K, as shown in Figure 1.1. It can be seen that the Seebeck coefficient of $Mg_{3.2}Sb_{0.49}Bi_{1.5}Te_{0.01}$ crosses over from negative values (n-type) to positive values (p-type) at 773 K, which has been attributed to the Mg loss notable by microstructural analysis and results in the formation of Mg vacancies that pin the Fermi level, eventually enabling the p-type conduction. Fortunately, coating the surfaces of the $Mg_3Sb_{2-x}Bi_x$ alloys with boron nitride was found to significantly improve their thermal stability by effectively limiting Mg loss at high temperatures. The time-dependent Hall carrier concentration of $Mg_3Sb_{2-x}Bi_x$ with cationic doping by La (Imasato et al. 2018), Yb (Wood et al. 2020) and Y (Shi et al. 2019c) at high temperatures and in a dynamic vacuum was also explored, and the cation-doped samples showed a delayed decrease in the carrier concentration compared with their Te-doped counterparts, indicating that doping with a cation rather than with Te can also improve the thermal stability of $Mg_3Sb_{2-x}Bi_x$ to some extent. Considering the large differences in the vapor pressures of the elements in $Mg_3Sb_{2-x}Bi_x$, high-temperature Mg loss that results in structural instability and decreased thermoelectric performance is common, thus restricting the operating temperature ranges of these materials. This disadvantage overshadows the high-temperature zT values and makes practical applications less appealing at temperatures higher than 673 K. Therefore, determining how to improve the thermal stability of $Mg_3Sb_{2-x}Bi_x$ compounds is more significant than simply focusing on enhancing their zT values.

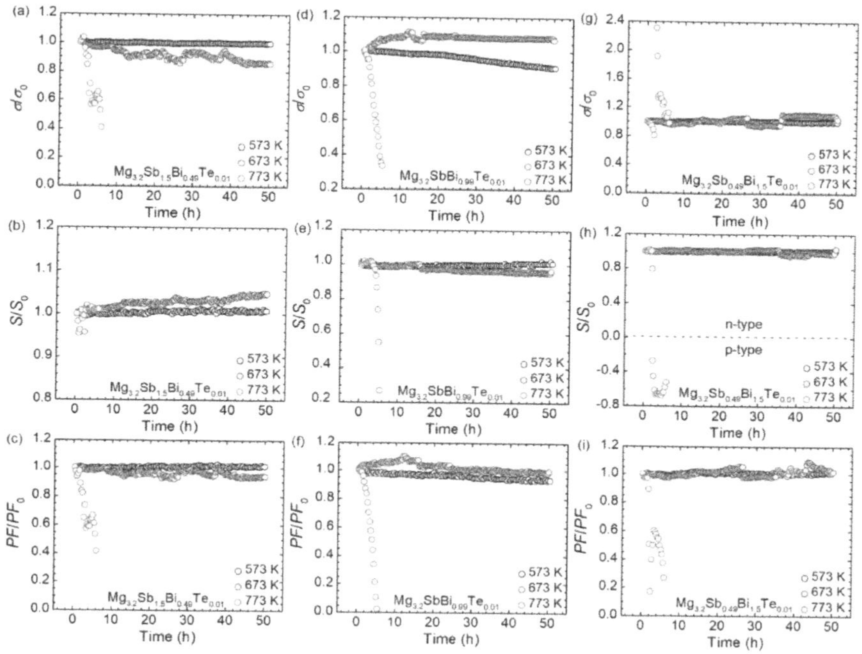

Figure 1.1. *Long-term in situ measurement of thermoelectric properties of $Mg_3Sb_{2-x}Bi_x$ alloys at 573 K, 673 K and 773 K. (a)–(c) σ/σ_0, (d)–(f) S/S_0 and (g)–(i) PF/PF_0 for $Mg_{3.2}Sb_{1.5}Bi_{0.49}Te_{0.01}$, $Mg_{3.2}SbBi_{0.99}Te_{0.01}$ and $Mg_{3.2}Sb_{0.49}Bi_{1.5}Te_{0.01}$, respectively. σ_0, S_0 and PF_0 represent the initial values; σ, S and PF represent the measured values at the specified time. Reprinted from Shang et al. (2020b). For a color version of this figure, see www.iste.co.uk/akinaga/thermoelectric2.zip*

1.4. Zn_4Sb_3

β-Zn_4Sb_3 is a well-known p-type thermoelectric material for applications in the intermediate temperature range (473~673 K) and exhibits high zT values up to 1.3 at 673 K (Zou et al. 2015). The high thermoelectric performance of β-Zn_4Sb_3 and its good potential for thermoelectric applications mainly result from its low lattice thermal conductivity (~0.3 W m^{-1} K^{-1} above 500 K), which is largely caused by its disordered structure that includes interstitial Zn (Dasgupta et al. 2013).

However, with the increasing temperature, β-Zn_4Sb_3 starts to decompose at 473 K due to a loss of Zn, forming ZnSb and Zn as decomposition productions and degrading its thermoelectric performance (Hao et al. 2011; Yang et al. 2017a, 2017b). Thus, it is of great significance to improve the thermal stability of β-Zn_4Sb_3 for practical applications. Based on both experimental results and theoretical calculations, doping with heavier metals can hamper the Zn migration and subsequently freeze β-Zn_4Sb_3 in its local disordered structure, eventually improving its thermal stability. It has also been found that this doping strategy can optimize the compound's electrical properties. For example, Wang et al. reported that Pb-doped β-Zn_4Sb_3 shows increased carrier mobility, achieving an optimized zT value of ~1.2 at 660 K (Wang et al. 2012), while its thermal stability is clearly improved due to the decreased Zn evaporation at high temperatures. Similar results were also reported by Deng et al. (2017). Furthermore, the addition of any of Ag (Song et al. 2018), Cd (Yin et al. 2010), Ge (Moghaddam et al. 2019), In/Sn (Shai et al. 2015), ZnO (Tang et al. 2015), etc. into β-Zn_4Sb_3 can also simultaneously improve its thermoelectric performance and high-temperature stability.

1.5. Skutterudites

In recent years, skutterudite-based materials have attracted significant attention due to their outstanding performance over a wide temperature range and their relatively inexpensive constituent elements. Among the various kinds of skutterudites, $CoSb_3$, a diamagnetic (Co^{3+}, $3d^6$) narrow-bandgap semiconductor, demonstrates high carrier mobility and a relatively large effective electron mass. Generally, $CoSb_3$-based thermoelectric generators can operate at temperatures up to 773 K. However, a series of studies have revealed that $CoSb_3$-based skutterudites undergo oxidation and Sb sublimation when the operating temperature exceeds 673 K, resulting in short service life (Sales et al. 1996; Zhang et al. 2016; Liu et al. 2020). Therefore, effective and reliable approaches are needed to address these instability issues.

The most effective method reported thus far is physical isolation, or inhibiting the sublimation of elements by spraying on a protective coating such as metal (El-Genk et al. 2006), ceramic (Zhang et al. 2016; Chen et al. 2013), glass (Dong et al. 2012), aerogel (Sakamoto et al. 2008) or a multilayered composite (Xia et al. 2014; Zhang et al. 2016). Introduction of the coating under certain conditions can indeed improve the stability of $CoSb_3$-based skutterudites at high temperatures, but it may result in some new problems at the same time. For example, the diffusion of a metal coating into the thermoelectric legs may result in decreasing performance over increasing thermal cycles (Dong et al. 2012). Moreover, the metal coating would cause an electrical short in actual applications. Although aerogel coating is highly robust and resistant to high temperatures, it could include micropores that may serve as escape channels for elemental oxidation and sublimation. Coating with another material such as glass or a ceramic may result in a mismatch of coefficient of thermal expansion (CTE) values or poor adhesion strength. Thus, coating with a multilayered structure consisting of different kinds of material is proposed. Zhang et al. (2016) reported a more effective multilayered coating with Mo, SiO_x and glass-ceramic layers. As shown in Figure 1.2, compared with samples without coating and with single-layered coating, that with the multilayered coating exhibits better stability in air at 873 K, indicating the effective suppression of both oxidation and sublimation. Nevertheless, suppressing decomposition in and improving the stability of skutterudite-based materials remain very challenging in practical application scenarios.

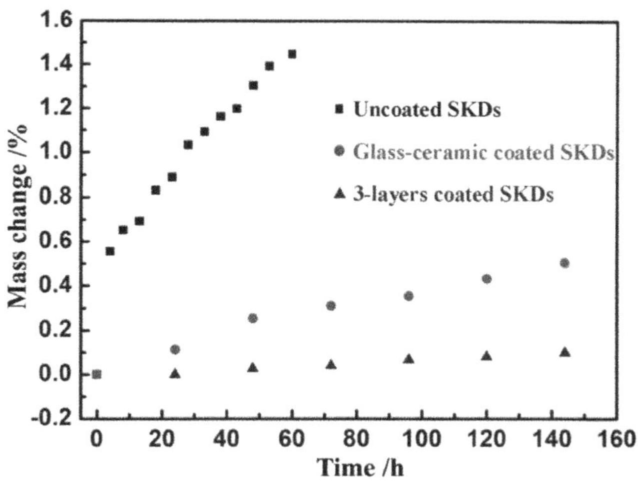

Figure 1.2. *Mass change of coated and uncoated skutterudites at 873 K in air as a function of time. Reprinted from Zhang et al. (2016). For a color version of this figure, see www.iste.co.uk/akinaga/thermoelectric2.zip*

1.6. $Cu_{2-x}X$ (X = S, Se, Te)

Copper chalcogenides $Cu_{2-x}X$ (X = S, Se, Te) possess complex atomic arrangements, and the Cu-deficient region in the Cu–Se system ($Cu_{2-x}Se$)

exhibits two distinct phases, a low-temperature α phase and an intermediate-temperature β phase (Liu et al. 2012). The low-temperature α phase with a low-symmetry crystal structure, where the Cu atoms are ordered and not superionic, is stable up to ~400 K and eventually transitions to the intermediate-temperature β phase with a cubic structure at ~400 K. In the β phase, the superionic Cu ions are kinetically disordered, leading to its remarkably low thermal conductivity and high thermoelectric performance. Significant advances have recently been obtained in studies of the p-type $Cu_{2-x}Se$ materials, and the zT value reportedly reaches 2.4 at 1,000 K for $Cu_{2-x}Se$/carbon nanotube (CNT) hybrid materials, showing promise for applications (Nunn et al. 2017, Lu et al. 2020b, Zhang et al. 2020b).

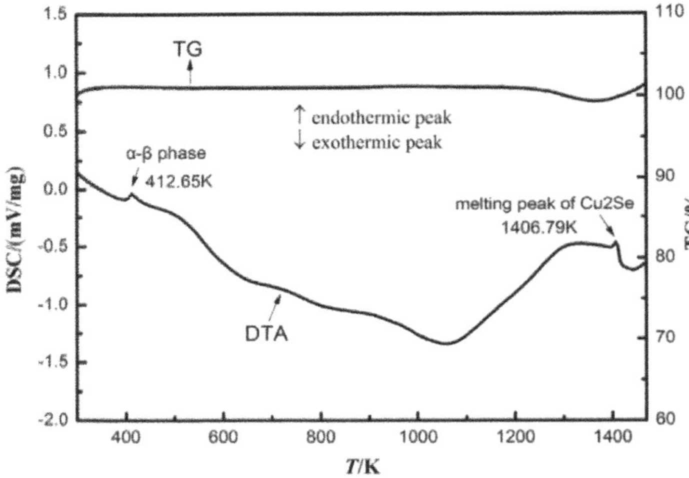

Figure 1.3. *TG-DTA curves for single-crystal Cu_2Se with a heating rate of 10 K/min. Reprinted from Liu et al. (2019)*

Unfortunately, $Cu_{2-x}Se$ materials suffer from thermal instability issues when they are applied at a high temperature and exposed to a large current, limiting their application in large-scale thermoelectric power generation (Shi et al. 2020). Thus, it is urgent to find out the mechanism of the instability issues of $Cu_{2-x}Se$ materials and improve their performance liability. Bohra et al. (2016) studied the effects of thermal cycles on the thermoelectric properties of $Cu_{2-x}Se$ in the temperature range of 320~1,070 K and found that electrical resistance increased with increasing thermal cycles, resulting in a reduction in the power factor from 13 μW cm^{-1} K^{-2} for the first cycle to

8 μW cm^{-1} K^{-2} for the third cycle at ~1,070 K. This change was attributed to Se loss and phase transition with increasing thermal cycles. Xue et al. (2019) reported that a high synthesis temperature not only enhances the thermoelectric performance of $Cu_{2-x}Se$ but also effectively improves its thermal stability; specifically, the Seebeck coefficient, electrical conductivity and thermal conductivity were found to be nearly constant between the heating and cooling cycles over the temperature range of 300~723 K for a sample fabricated at 1,273 K. Furthermore, a Ni/Mo/$Cu_{1.96}Se_{0.8}S_{0.2}$ thermoelectric unileg was able to exhibit a very steady power output over 400 thermal cycles between 473 K and 873 K through the control of its chemical composition (Mao et al. 2019b). Other dopants, such as Li (Kang et al. 2017), CNTs (Nunn et al. 2017), W (Bohra et al. 2020) and Cd (Lu et al. 2020a), can also improve the thermal stability of $Cu_{2-x}Se$ materials, allowing them to be applied for power generation. Additionally, single-crystal α-$Cu_{2-x}Se$ was fabricated using the Bi-flux method and no weight loss occurred below the melting point (~1,403 K), as shown by the thermogravimetric-differential thermal analysis (TG-DTA) results displayed in Figure 1.3. The DTA curve shows the endothermic peak at ~400 K, which has been attributed to the Cu_2Se α–β phase transformation. The power factor of the single-crystal samples did not obviously decrease after three thermal cycles in the temperature range of 300~690 K, demonstrating their improved thermal stability compared with the polycrystalline materials (Liu et al. 2019). However, single crystals are usually more expensive than the polycrystalline materials, so they are not preferred for large-scale applications.

1.7. GeTe

GeTe-based thermoelectric materials were first studied in the 1960s, and they re-emerged as a subject of research interest only recently. Thanks to band engineering and defect engineering strategies, the zT values exceeding 2 have been achieved in GeTe-based materials, showing their promise for applications (Hong et al. 2020; Zhang et al. 2020a). Nevertheless, the actual application of GeTe-based modules have a major obstacle to overcome concerning how the pristine GeTe undergoes a reversible transition between a low-temperature rhombohedral structure and a high-temperature cubic structure around 700 K, as illustrated in Figure 1.4. It should be noted that such structural instability can seriously reduce the reliability of devices if it is not handled properly. Additionally, there is a considerable change in the coefficient of thermal expansion (CTE) between the two phases

(11.2×10^{-6} K^{-1} for the rhombohedral structure and 23.4×10^{-6} K^{-1} for the cubic structure), and such CTE change is extremely detrimental in device applications since it can cause significant thermal stress during operation (Xing et al. 2021). Thus, methods to suppress the phase transition are required.

Fortunately, Liu et al. (2018) successfully obtained GeTe-based material that is in the cubic phase over the entire operating temperature range through simple Mn and Bi co-doping at the Ge site. Soon afterwards, Zheng et al. (2018) reported that the lower-temperature cubic GeTe can also be achieved by Mn and Sb co-doping and found that the transition temperature was reduced from 700 K closer to room temperature. The above two studies show that Mn plays a significant role in inhibiting the phase transition. Bi and Sb can also adjust the crystal structure to a certain extent, but they are more responsible for regulating the carrier concentration. Recently, resonant bonding properties induced by a symmetrized crystal lattice were found in the Ti-doped GeTe-based material (Li et al. 2020c). Ti doping not only can reduce the lattice constant c/a ratio to achieve a cubic structure but can also enhance the Seebeck coefficient due to the increased band degeneracy (Shuai et al. 2019).

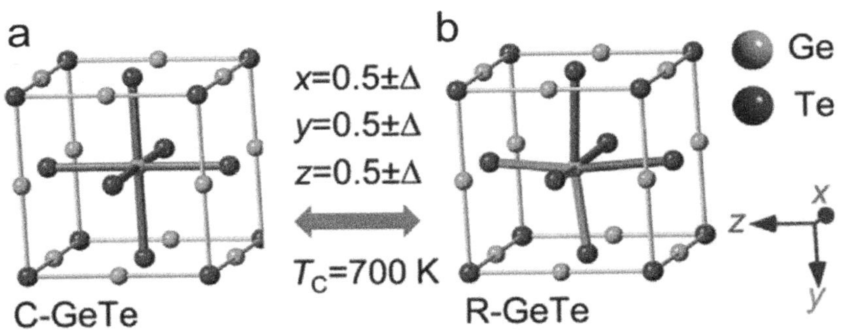

Figure 1.4. *Structure of GeTe: (a) cubic and (b) rhombohedral. Reprinted from Li et al. (2021). For a color version of this figure, see www.iste.co.uk/akinaga/thermoelectric2.zip*

1.8. Outlook on thermoelectric materials stability

Admittedly, the instability issue is universal in thermoelectric materials because of their weak chemical bonding and highly disordered content. In this section, different typical instability issues faced by several kinds of materials were briefly introduced, and the corresponding strategies to address these issues were also discussed in detail. In addition to the above-discussed thermoelectric materials, other promising materials with excellent thermoelectric performance, including PbTe (Wang et al. 2018), BiCuSeO (Li et al. 2014) and half-Heuslers (Rausch et al. 2016), also suffer from thermal instability, which markedly impedes their commercial applications, especially for the generation of thermoelectric power. Generally, the instability issue can be solved either chemically or physically. From a chemical point of view, doping or heat treatment can improve the stability of a thermoelectric material's constituent elements and structure on a microscopic level. For some special materials with phase transitions, reasonable chemical doping can alter the phase-transition point, thereby avoiding the instability of the material in the operating temperature range. From a physical point of view, isolating a material from the external environment, such as with a coating or through vacuum packaging, is also an effective measure to effectively enhance its stability. These strategies should be thoroughly evaluated and rationalized in future studies with the help of advanced characterization tools and theoretical simulations. Without these methods, it would be difficult to understand the origin of phase-transition regulation or structure-stability preservation when introducing external dopants. In addition, for long-term stability, developing a rapid and scalable method for preparing thermoelectric materials with fewer defects is advantageous. In short, to achieve the practical uses of thermoelectric materials, greater efforts should be made with enhancing the thermal stability of existing materials and searching for new materials with both high performance and good stability. In particular, reporting the stability data of high-zT materials together with their transport property measurements is recommended for at least the papers on device performance.

1.9. Thermoelectric device design analysis

1.9.1. *Thermal stress analysis*

Despite the tremendous improvements in thermoelectric material performance over the last several decades, research into building thermally and mechanically reliable devices has been very slow. Since thermoelectric modules are subjected to large temperature gradients over long periods, the thermomechanical properties and thermal stability of various constituent components must be considered in device design and manufacturing (Karri et al. 2018). Significant thermal stress, deformation and even failure may occur in thermoelectric devices due to mismatches in the thermal, mechanical and rheological properties of the constituent components caused by material specificity, cyclic thermal loading and possible mechanical vibration (Music et al. 2016). As a result, in designing a thermoelectric device, it is critical to conduct temperature distribution and thermal stress analyses in order to effectively reduce thermal stress through material selection or structure optimization. The main aim of such a thermomechanical stress study is to maintain the thermoelectric device's high output performance under a long-term operation.

In this section, thermomechanical stress in thermoelectric modules is discussed in depth. We first discuss the coefficient of thermal expansion, which is a significant factor in determining thermal stress, as well as other design considerations relevant to thermal stress, including module configuration and geometry. We then present an analysis of the lowest value of thermal conductivity based on the maximum allowable thermal shear stress, emphasizing the importance of balancing the high zT generated by phonon scattering with mechanical reliability. These analyses elucidate the relationship between the above factors and the performance, reliability and durability of thermoelectric devices, shedding light on the optimal design of reliable thermoelectric devices.

1.9.1.1. *Coefficient of thermal expansion (CTE)*

The coefficient of thermal expansion (CTE, α) of a material is its length change in response to temperature variation. Differing expansion of constituent parts of a thermoelectric device under a considerable temperature gradient would result in high thermal stress and hence a greater risk of interfacial damage during long-term operation (He et al. 2018). For example, cracking or exfoliation at the interface of an oxide-based module is due to the widely differing CTEs of oxides and metallic contact materials (Liu et al. 2015).

The temperature-gradient-induced mechanical stress $\xi(T)$ of a thermoelectric material can be approximated by:

$$\xi(T) = \frac{\alpha(T)E(T)\Delta T}{(1 - \nu(T))} \quad [1.1]$$

where α, E and ν are its CTE, Young's modulus and Poisson's ratio, respectively (Ni et al. 2013).

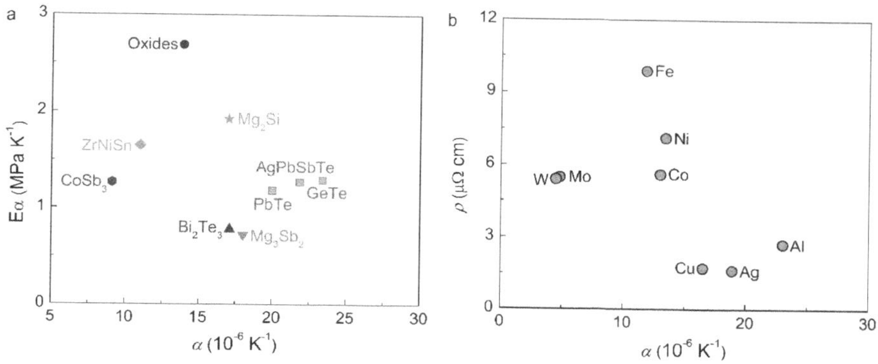

Figure 1.5. *(a) Coefficient of thermal expansion (CTE, α) and elastic modulus-CTE product (Eα) for representative thermoelectric materials. Data obtained from Ni et al. (2013), Zhao and Tan (2014) and Music et al. (2016). (b) CTE and electrical conductivity (r) for pure metals. Data obtained from Chowdhury et al. (2009). For a color version of this figure, see www.iste.co.uk/akinaga/thermoelectric2.zip*

Clearly, a large CTE value and a high Young's modulus are undesirable for reduced thermal stress. According to calculation results, a CTE mismatch between different components in a thermoelectric module can result in excessive thermal stress and in situ fracture (Ravi et al. 2009). In other words, CTE has a significant impact on the module's thermomechanical robustness. Hence, obtaining thermal expansion statistics for commonly used thermoelectric materials and investigating the underlying relationship between the structure and the thermal expansion coefficient of each are critical for achieving thermally match systems in thermoelectric devices. The measured CTE results for commonly investigated thermoelectric materials with high zT values are summarized in Figure 1.5a. Clearly, most high-performance thermoelectrics have CTE values of more than 15×10^{-6} K^{-1}, which can be attributed to the weak chemical bonding in these materials that also results in their low lattice thermal conductivity. The large CTE values emphasize the importance of selecting appropriate metal electrodes and designing reliable contact layers for the thermoelectric legs in a module in order to reduce thermal stress (Ni et al. 2013). Additionally, the CTE values of thermoelectric compounds should be optimized to establish better matches between p- and n-type materials or between thermoelectric materials and electrodes by tuning the compositions, filler atoms, grain sizes (Rogl et al. 2017) and additives (Rogl et al. 2010). However, these solutions need to be carefully examined to ensure that high zT values are retained. The CTE and electrical conductivity values of pure metals are shown in Figure 1.5b as a reference to choose suitable contact layers for both CTE matching and low electrical resistivity. Music et al. (2016) used the first-principles calculation method to investigate the thermomechanical response of 20 thermoelectrics. They found that oxides exhibit larger products of linear thermal expansion coefficient and elastic modulus, implying that thermoelectric oxides are susceptible to thermally induced stress, shock and fatigue. For convenience in prediction and matching, it was also suggested that equilibrium volume, an easily measurable parameter, can be used to estimate this product and the thermal response of thermoelectric materials.

1.9.1.2. *Module configuration and geometry design*

Realizing high conversion efficiency or output power density in a thermoelectric module relies on a specific configuration and geometry, so its fabrication always involves the design and optimization of these factors (Wu et al. 2014). The principle of thermoelectric device configuration design is to minimize the internal stress caused by temperature difference and thermal shock while making full use of the heat source.

The π-shaped, O-shaped and Y-shaped structures, schematically illustrated in Figure 1.6, are the most common thermoelectric-module configurations (Zhang et al. 2016). In a typical π-shaped module as shown in Figure 1.6a, many thermoelectric unicouples consisting of an n-type leg and a p-type leg are connected electrically in series and thermally in parallel. In order to maximize the use of temperature differences, the direction of heat flow is usually perpendicular to the substrate. It should be noted that when this design is applied at higher temperatures, the thermally induced stress, expansion and diffusion of different materials should be thoroughly considered because the repeated heat shock may cause severe deformation and even device failure, especially at the contacts. An O-shaped module (Figure 1.6b) consists of a coaxial assembly of many annular flat gaskets consisting of alternating n- and p-type thermoelectric rings. Compared to the π-shaped configuration, the O-shaped configuration makes it easier to achieve device integration, efficient heat transfer and low-stress structure design, but the manufacturing and joining technology needed for the specially shaped thermoelectric elements and the metal electrodes is very difficult to master if not impossible. The Y-shaped module configuration (Figure 1.6c) has been designed mainly to eliminate the impact of thermal expansion mismatch between different constituent components while maintaining excellent thermoelectric performance. In this configuration, rectangular n- and p-type thermoelectric materials are alternately connected to electrode plates in a sandwich structure, which allows the thermoelectric elements to be maintained in a non-fully constrained state and reduces the thermal expansion and deformation caused by the constraints of the hot and cold sides. The horizontal series connection of the n- and p-type materials also avoids the stress concentration caused by the difference in their thermal expansion coefficients.

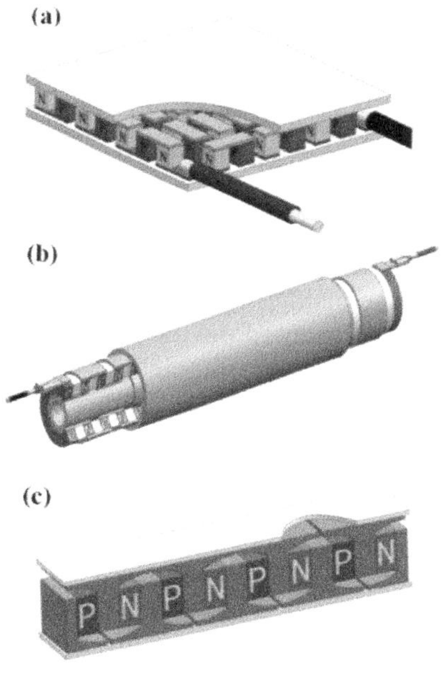

Figure 1.6. *Schematic diagrams of (a) π-shaped, (b) O-shaped and (c) Y-shaped modules. Reprinted from Zhang et al. (2016). For a color version of this figure, see www.iste.co.uk/akinaga/thermoelectric2.zip*

In addition to the shape of the module, the unicouple geometry plays an important role in maximizing performance and maintaining it under large thermal stress (Shittu et al. 2020). Recent thermodynamic and thermal stress studies of thermoelectric generators have revealed that the highest stress level is found at the edges of a module, particularly those on its hot side (Al-Merbati et al. 2013). By tuning the pin geometry, the temperature variation can be improved, and thus the lifetime of the module can be prolonged. Fabián-Mijangos et al. (2017) compared the effects of various leg shapes, including square, hexagon, octagon and circular prisms, on overall thermal stress and found that cylindrical legs are preferable under the fixed hot side condition, while square prisms are more reliable under the free hot side condition. Although modifying the leg dimensions can improve a module's thermal reliability, a unilateral focus on thermodynamic properties may result in low output-power performance. Erturun et al. (2015) found that there is an inverse relationship between output performance and thermal integrity by using three-dimensional finite-element analysis and experiments, indicating the importance of determining the optimal shape for balancing performance and stability. More specifically, Sarhadi et al. (2015) recommended that the thermoelectric legs should be long enough to achieve high efficiency and low thermal stress while short enough to maintain a high output power density. A segmented module presents a more complicated situation since more factors are involved in its structural optimization. In most cases, a compatibility factor match between segmented materials is sufficient to produce an effective segmented device. However, Jia et al. (2014) noted that when considering the strength requirements, a high conversion efficiency in a segmented device is not always possible. In short, multi-parameter optimization of thermoelectric generators is required to achieve both maximum efficiency and structural reliability.

Modulating the hot-side design of a thermoelectric device by taking into consideration the connection structure and heat flux direction is another strategy to reduce thermal stress and obtain a durable interface. Sun et al. (2021) examined a new design that uses a step structure of ceramic plates with thermoelectric legs inserted into them, which separates the heat and current conduction paths at the hot-side electrode (Figure 1.7). As a result, interconnect cracking is only affected by the connection between the hot-side ceramic plate and the metallic interconnect adjacent to it, which reduces the thermal stress damage of the interconnect and increases the number of operation cycles prior to failure. Additionally, the design was calculated to have a favorable impact on the module's power output performance, showing simultaneous improvement in thermomechanical stability and thermoelectric performance. In the future, a real module should be prepared and evaluated to substantiate this new design and initiate further improvements, and also to show the cost difference since the new design will be more expensive.

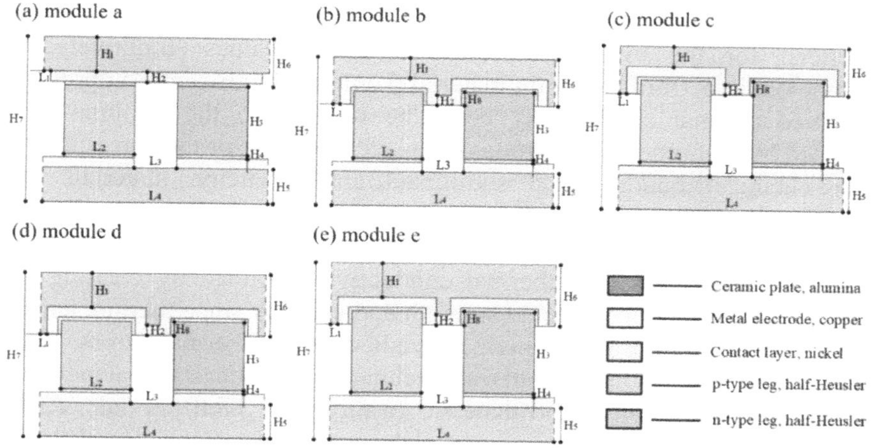

Figure 1.7. *Schematic diagrams of thermoelectric modules featuring (a) the conventional thermoelectric unicouple geometry and (b–e) the new hot-side design geometry with different parameters. Reprinted from Sun et al. (2021). For a color version of this figure, see www.iste.co.uk/akinaga/thermoelectric2.zip*

When considering the substrate design specifically, there are two main strategies for protecting the module from thermal stress. Increasing the thickness of the ceramic plates can relieve thermal stress while ensuring that the device's output performance remains nearly constant. Another effective way to reduce thermal stress is to replace rigid substrates with compliant pads for electrical insulation (Kambe et al. 2010). Nevertheless, in this case, the derived module will be mechanically weak and will need additional encapsulation. Most thermoelectric modules currently investigated use alumina substrates because of their high thermal conductivity. A Cu electrode strip can be deposited onto each substrate prior to assembly. The thickness of the Cu electrode strips can be adjusted according to the maximum expected output current of the module.

1.9.1.3. *Materials design for reduced thermal stress: engineering thermal conductivity κ_{eng}*

Decreasing lattice thermal conductivity by introducing multi-scale microstructures has been widely applied to improve the performance of thermoelectric materials. In particular, nanostructured grains are proposed to scatter phonons significantly without degrading electrical transport, which is beneficial for the overall improvement of zT values. Intuitively, the thermoelectric performance of a material would be significantly improved if its lattice thermal conductivity could be reduced to the minimal value predicted by theoretical calculations. However, because of the corresponding weak chemical bonding and significant anharmonicity, materials with exceptionally low lattice thermal conductivity are usually brittle, limiting their practical applications. In other words, even though carrier mobility is unaffected, the low lattice thermal conductivity attained by creating vast defects may degrade mechanical strength and stability since thermoelements will be subjected to large thermal gradients and thermal stress. Thus, reconsidering the phonon scattering technique is critical to balancing zT values and mechanical brittleness. Inspiringly, Kim et al. (2016) hypothesized that when exposing materials to shear thermal stress, there is a minimal value of thermal conductivity which is safe to employ (Figure 1.8). To begin, they proposed that there is an appropriate range of leg length (L), where the lower limit of leg length (L_{min}) is dictated by the yield strength and the upper limit of leg length (L_{max}) may be computed using thermoelectric material parameters and boundary conditions. For materials with higher thermal conductivity, a larger value of L_{max} could be expected for a fixed cold-side temperature. If L_{max} is greater than L_{min}, it is safe to reduce the thermal conductivity so that L_{max} is closer to L_{min}, therefore enhancing the efficiency of the device. Engineering thermal conductivity (κ_{eng}) is the thermal conductivity value when L_{min} and L_{max} are the same. Thus, κ_{eng} can be defined as the lower limit of a material's thermal conductivity that still assures the device's thermal reliability in a steady-state operation. On the other hand, since thermoelectric devices are subjected to thermal loading and thermal shock in real applications, it is also important to investigate the minimum thermal conductivity of thermoelectric materials during transient operation, which can be estimated from the dynamic boundary conditions and the thermomechanical properties of the materials. Based on these findings, Kim et al. (2016) investigated a Bi_2Te_3-based module as a case study and found that further reduction in thermal conductivity for zT improvement should not be attempted because the thermal conductivity of current-generation p-type Bi_2Te_3 already approaches the steady-state κ_{eng}. In contrast, more reliable performance improvement may rely on boosting its electrical properties.

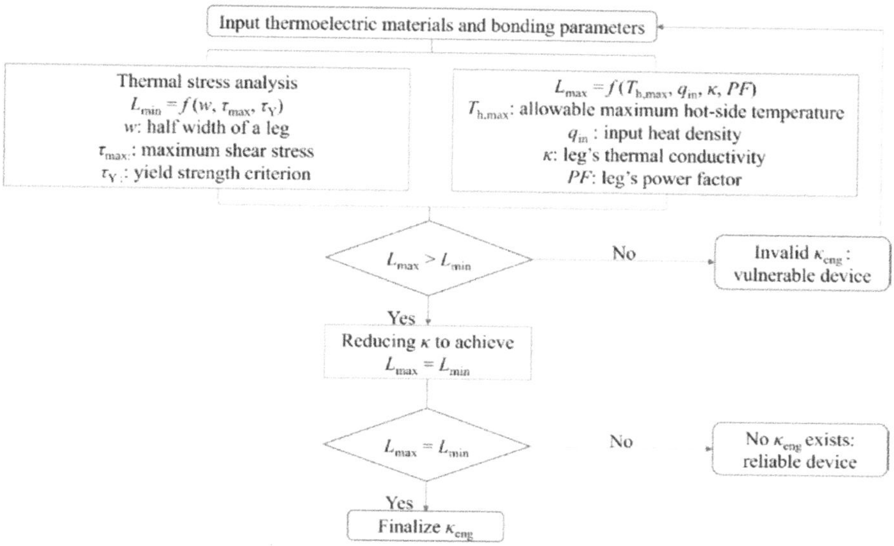

Figure 1.8. *Flow chart for calculating engineering thermal conductivity (κ_{eng}) in the steady state. Reproduced from Kim et al. (2017)*

1.9.2. *Interface analysis, design and fabrication*

The interfaces between thermoelectric materials and metal electrodes in a thermoelectric device undergo complex physical and chemical processes during operation due to repeated thermal cycling and thermal shock. In addition to the thermal stress mentioned above, more engineering problems, such as bonding strength, contact resistance and high-temperature diffusion, are confronted when considering interfaces. These issues have a great impact on the reliability and durability of the device. In the following section, the relationship between the interface properties and the device performance, as well as the failure behavior of the interfaces are first analyzed in detail, demonstrating the great importance of the interfaces. Subsequently, general selection and design principles for high-performance and reliable interfaces are introduced.

1.9.2.1. *Interface-induced performance degradation and device failure*

Contact thermal and electrical resistance at the interfaces between different components of the module will result in lower power output performance because of the wasted electricity and heat. Such a performance discrepancy can occur at either a pristine imperfect joining or an interface that is increasingly degraded over long-term thermal cycles. In order to evaluate the effect of the interface contacts on device performance, Kim (2018) developed a numerical method for the performance analysis of thermoelectric generators with thermal and electrical contact resistance, in which the effectiveness was defined as the ratio of the thermoelectric generator's power output with any resistance to that without resistance. Effectiveness thus indicates how much of the potential performance is actually achieved in the presence of any resistance, and the corresponding mathematical results and numerical modeling for this parameter are shown in Figure 1.9. It is clear that the effectiveness increases with increasing thermal contact conductance (i.e. decreasing thermal contact resistance) and decreases with increasing electrical contact resistivity. According to the numerical simulation, low contact thermal conductance can cause performance degradation of up to ~30%, while large electrical contact resistivity can cause performance degradation of more than 40%. The results of this study emphasize the importance of interfaces on the actual performance of thermoelectric generators.

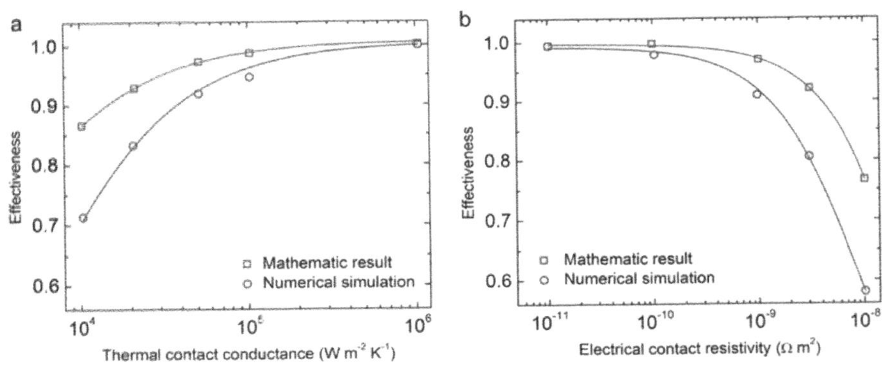

Figure 1.9. *Effect of (a) thermal contact conductance and (b) electrical contact resistivity on thermoelectric device performance. Data obtained from Kim (2018). For a color version of this figure, see www.iste.co.uk/akinaga/thermoelectric2.zip*

For thermoelectric cooling devices, the theoretical value of the coefficient of performance (COP) without considering thermal or electrical contact can be obtained by the formula:

$$\mathrm{COP} = \frac{T_c}{T_h - T_c}\frac{\sqrt{1+\overline{ZT}} - \frac{T_h}{T_c}}{\sqrt{1+\overline{ZT}} + 1}. \qquad [1.2]$$

This formula is based on the theory developed by Ioffe in 1957 and is not highly accurate in evaluating thermoelectric modules (Ioffe 1957). In 1996, Rowe and Min proposed a more complete method, in which the thermal and electrical contacts are considered (Rowe 1995). The output voltage V and current I can be expressed by:

$$V = \frac{NS(T_h - T_c)}{1 + \frac{2\kappa l_c}{\kappa_c l}}, \qquad [1.3]$$

and

$$I = \frac{AS(T_h - T_c)}{2\rho(\frac{2\rho_c}{\rho} + l)(1 + \frac{2\kappa l_c}{\kappa_c l})} \qquad [1.4]$$

where N, S, A, l, l_c, κ, κ_c, ρ and ρ_c are the number of thermocouples in the module, the Seebeck coefficient, the cross-sectional area of the thermoelectric material, the length of the thermoelectric leg, the thickness of contact layer, the thermal conductivity of the thermoelectric leg and of the contact layer and the electrical resistivity of the thermoelectric leg and of the contact layer, respectively. $\kappa_{ratio} = \kappa/\kappa_c$ and $\rho_{ratio} = 2\rho_c/\rho$ are standard thermal and electrical contact parameters, respectively. Generally, $\kappa_{ratio} \sim 0.2$ and $\rho_{ratio} \sim 0.1$ mm are appropriate values for commercially available Peltier modules. The modified conversion efficiency $\mathrm{COP_{rev}}$ can be derived as:

$$\mathrm{COP_{rev}} = \frac{(T_h - T_c)/T_h}{(l + 2\kappa_{ratio}\frac{l_c}{l})^2\{(2 - \frac{1}{2}[(T_h - T_c)/T_h] + (4/ZT_h)[(1+\rho_{ratio})/(1+2\kappa_{ratio} l_c)]\}}, \qquad [1.5]$$

where $Z = S^2/\rho\lambda$. COP_{rev} could be further derived for operation in the Peltier mode:

$$COP_{rev} = \frac{l}{l + 2\kappa_{ratio} l_c} \left(\frac{T_c}{T_h - T_c} \frac{\left[1 + \frac{Z\overline{T}l}{\rho_{ratio} + l} \right]^{1/2} - \frac{T_h}{T_c}}{\left[1 + \frac{Z\overline{T}l}{\rho_{ratio} + l} \right]^{1/2} + 1} - \frac{\kappa_{ratio} l_c}{l} \right), \quad [1.6]$$

where $\overline{T}$ is the mean temperature across the thermoelectric module. Based on this formula, the relationship between COP_{rev} and the length of the thermoelectric material at different κ_{ratio} and ρ_{ratio} values is shown in Figure 1.10 for $l_c = 0.5$ mm, $Z = 2\times10^{-3}$ K^{-1} and $\Delta T = 25$ K with T_h at 300 K. It is clear that COP_{rev} varies under different κ_{ratio} and ρ_{ratio} values. The COP_{rev} of a 1-mm leg increases by ~18.6% when κ_{ratio} decreases from 0.1 to 0.001 (i.e. the thermal conductivity of the contact layer increases), while it decreases by ~12.6% when ρ_{ratio} increases from 0.1 mm to 0.2 mm (i.e. the electrical resistivity of the contact layer increases). A dramatic decrease can be observed in COP_{rev} when the length of the thermoelectric leg is less than 2 mm. The thermal and electrical resistance of the contact layer become increasingly more important once the length of the thermoelectric leg is relatively short. Therefore, in thermoelectric cooling devices, for which the service environment is not as harsh as that for the thermoelectric generators, a relatively slight increase in contact resistance still leads to notable degradation in the cooling performance.

In actual applications, device failure frequently occurs at the interfaces which undergo repeated thermal shocks, which place greater demands on the interfaces. Figure 1.11 shows the results of long-term thermal stability testing of a $MnSi_{1.77}/Mg_2Si_{1-x}Sn_x$-based thermoelectric module (Mejri et al. 2020) and a Bi_2Te_3-based thermoelectric leg (Barako et al. 2012). For the $MnSi_{1.77}/Mg_2Si_{1-x}Sn_x$-based thermoelectric generator, the thermoelectric legs were soldered to metalized ceramic plates using silver, as shown in Figure 1.11a. The barrier layer between the legs and the solder consists of a thin layer of Au/Ti (300/100 nm). During testing, thermal cycling between 423 K and 673 K was applied at the hot side of the module while its cold side was maintained at 323 K. After 400 cycles (~100 hours) of testing, serious cracks occurred at the hot-side material interface, as shown in Figure 1.11b. In another example, Bi_2Te_3-based thermoelectric legs were tested by quickly switching the temperature between 253 K and 419 K at one side for ~45,000 cycles, while the other side was maintained at the ambient temperature, and the results are shown in Figures 1.11c and 11d. It can be clearly seen that cracks developed at the interface between the thermoelectric leg and the electrode, which would undoubtedly cause performance degradation. In summary, whether due to thermal shock or large interface resistance, imperfections in the contact layer can have a great impact on device performance, suggesting the importance of interface research for reliable thermoelectric modules.

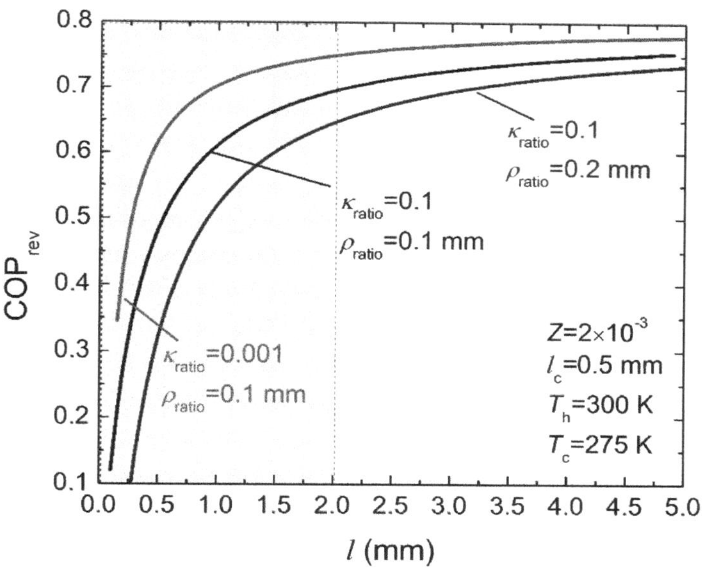

Figure 1.10. *COP_{rev} as a function of thermoelectric-leg length (l). Data obtained from Rowe (1995). For a color version of this figure, see www.iste.co.uk/akinaga/thermoelectric2.zip*

Figure 1.11. *$MnSi_{1.77}/Mg_2Si_{1-x}Sn_x$-based thermoelectric generator (a) before and (b) after thermal cycling between 423 K and 673 K for 100 h. Reprinted from Mejri et al. (2020). Comparison of the interface between a Cu electrode and a Bi_2Te_3 leg (c) before and (d) after thermal cycling between 253 K and 419 K for ~45,000 cycles. Reprinted from Barako et al. (2012). For a color version of this figure, see www.iste.co.uk/akinaga/thermoelectric2.zip*

1.9.2.2. *Interface design principles*

A thermoelectric device can be reduced to its essential component, a complete circuit consisting of p- and n-type thermoelectric legs (i.e. a unicouple), as schematically illustrated in Figure 1.12. The contact interface of a unicouple is a complicated region containing many layers. The design of such a multi-layer interface usually requires a more complex processing technique. Ideally, the interface design should be as simple as possible since, in practical applications, a simple interface structure facilitates manufacturing and processing. Therefore, in early studies on thermoelectric devices, researchers attempted to connect the thermoelectric legs directly to a conductive-strip (e.g. Cu) substrate. However, numerous experimental results showed that the loss of thermoelectric energy conversion efficiency could be as high as 40% due to the large contact resistance caused by the diffusion between the bonding material and the thermoelectric legs (Ren et al. 2018). Additionally, this diffusion phenomenon also leads to performance deterioration in the thermoelectric legs, thereby shortening the service life of the device. Taking the Bi_2Te_3-based thermoelectric cooling devices as a typical case, it was found that although the Bi_2Te_3 legs and Cu strips can be joined at 503 K by using a Sn–Sb–Bi solder, the diffusion of Sn into the thermoelectric legs leads to high electrical resistance. Although the electrical resistance was reduced in later studies by replacing the Sn–Sb–Bi solder with a Sb–Ag solder, it was found that Ag diffusion into the thermoelectric legs could also be detrimental to the device (Rosi et al. 1962). In addition, Cu is also highly diffusible into Bi_2Te_3 at the operating temperature of ~523 K (Yang et al. 2013). Thus, the concept of adding a metal contact layer to serve as both the electrode and the solder diffusion barrier was proposed. For example, a Ni-metalized layer applied in a Bi_2Te_3-based thermoelectric device can effectively prevent the diffusion of the Sn from the Sn–Sb–Bi solder into the thermoelectric legs (Lin et al. 2017).

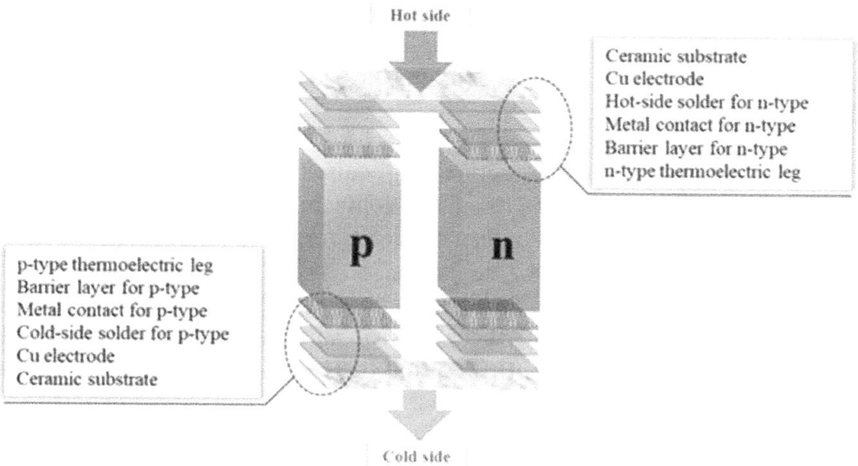

Figure 1.12. *Schematic diagram of the structure design of a thermoelectric unicouple. For a color version of this figure, see www.iste.co.uk/akinaga/thermoelectric2.zip*

A good metal contact layer and the corresponding connection technology should have the following characteristics for either a thermoelectric power generator or a thermoelectric cooler (He et al. 2018; Liu and Bai 2019): (1) the coefficient of thermal expansion (CTE) of the metal contact layer must match that of the thermoelectric leg; (2) the metalized layer must have high thermal conductivity and electrical conductivity to reduce extra energy loss at the interface; (3) low contact resistivity at the interface between the metal contact layer and the thermoelectric legs is necessary; (4) the connection technology between the metal electrodes and the thermoelectric legs should be simple, and the bonding strength should be high enough to ensure the stability of the device while operating at high temperatures and (5) avoiding doping or alloying effects due to high-temperature atomic diffusion at the interfaces is important since these could significantly alter the compositions of the thermoelectric legs and the electrode materials, leading to performance deterioration. The metal electrode should be inert to the thermoelectric material to avoid serious chemical reactions. Although diffusion is necessary for establishing a strong bond, it should be terminated when the joining is complete. In fact, it is very difficult to meet all of the above criteria at the same time. Therefore, determining how to balance various influencing factors in interface design is an important subject that needs to be studied for the development of thermoelectric devices. In general, the interface design needs to be adjusted according to actual application requirements.

From the perspective of a practical application, CTE matching is the most important issue when selecting the metal contact material since the mismatch of CTEs could cause severe internal stress, resulting in the failure of the device. Mismatched CTEs could also increase the difficulty in manufacturing thermoelectric devices. Ravi et al. (2009) carried out detailed measurements and analysis of the CTEs of some advanced thermoelectric materials, and their results confirmed that the CTE has a profound effect on the thermomechanical robustness of the devices in which these materials are incorporated. Therefore, CTE matching requires careful engineering to meet performance expectations such as long-term durability. The CTEs of various thermoelectric materials are quite different from one another, so the metal electrode materials require targeted optimization such as composition tuning in the contact layer that is sensitive to the elemental ratio of the thermoelectric legs (Liu and Bai 2019). Additionally, multi-layer contact materials can be applied to gradually match the CTE of the thermoelectric materials.

In addition to CTE matching, a strong bond between the metal contact layer and the thermoelectric legs is essential for device reliability and durability. Generally, the metal contact layer can be made through either a chemical route (i.e. electrochemical plating or electrochemical deposition) or a physical route (i.e. sputtering, spraying, or direct hot pressing) (Zhang et al. 2016). An interface made by electrochemical plating or electrochemical deposition is very thin, and although this reduces the total electrical and thermal resistance at the contact region, the bonding strength is relatively low (~10 MPa) (Weitzaman 1967; Feng et al. 2013). This bonding strength could be adequate for refrigeration applications like commercial Bi_2Te_3-based devices, but it is far lower than that needed for power generation under thousands of thermal cycles and thermal shocks. Many studies have confirmed that increasing the surface roughness through chemical etching or sandblasting could effectively enhance the bonding strength because it would result in a larger contact surface area and an enhanced anchor point effect (Zhang 2018; Liu and Bai 2019). Compared to the chemical methods for metal contact layer production, some physical techniques produce a stronger bond between a metal electrode and the thermoelectric legs; these include thermal spraying (e.g. 18 MPa~25 MPa bonding strength between a Ni-metalized layer and a Bi_2Te_3 thermoelectric leg) (Zhang et al. 2014) and direct hot pressing (e.g. 50 MPa bonding strength between a Ti-metalized layer and a half-Heusler thermoelectric leg) (Joshi and Poudel 2016). Unfortunately, greater bonding strength is accompanied by higher contact resistance since it is usually achieved by deep interdiffusion or interface reactions in which an intermetallic layer could be formed between the metal contact layer and the thermoelectric leg. Although moderate diffusion is beneficial to interface bonding strength, uncontrolled diffusion and chemical reactions may cause increased thermal and electrical resistance, as well as significant changes in the composition of the metal contact layer and the thermoelectric legs, which will eventually cause performance deterioration in the device, or even device failure. Therefore, a balance should be found between bonding strength and other factors such as diffusion rate and contact resistance.

Using a single-layer metal contact can simplify the production process, but it makes meeting all of the above requirements at the same time difficult in some cases. For instance, there may be a large mismatch between the CTEs of the metal electrode and the thermoelectric legs; the metal contact layer could react with the thermoelectric legs quickly to form an intermetallic layer, which may lead to the deterioration of the entire device; or it may not be possible to produce a strong bond between the metal contact layer and the thermoelectric leg. Therefore, multi-layer interface technology has been introduced (Xia et al. 2014; Zhang et al. 2016; Xing et al. 2021), in which a barrier layer (also known as a transition layer) between the electrode and the thermoelectric leg serves as a buffer not only to slow down the diffusion behavior but also to improve the wettability of the material in order to facilitate brazing. Meanwhile, the barrier layer itself should also have high thermal and electrical conductivity to reduce the extra energy loss at the interface and thus improve the reliability of the device.

1.9.2.3. *Joining*

Joining processes are of great significance for the interface properties of thermoelectric modules. In particular, a proper joining technology can reduce the interface stress and improve the bonding strength, thereby increasing the reliability of the thermoelectric device to a certain extent. Over the past decades, several methods, such as the chemical and physical routes mentioned above, have been applied to join metal electrodes and thermoelectric legs. Among these methods, two major ones stand out as the most reliable and suitable for industrial manufacturing: diffusion welding and thermal spraying (see Figure 1.13). Diffusion welding, also known as one-step sintering, has been successfully applied in the fabrication of $Mg_3Sb(Bi)_2$ Zintls, MgAgSb, skutterudites, half-Heuslers, etc. (Zhao et al. 2012; Joshi and Poudel 2016; Zhang et al. 2016; Yin et al. 2020; Ying et al. 2021). As shown in Figure 1.13a, for the diffusion welding process, the electrode material powder and the thermoelectric material powder (sometimes including barrier material powder) are added into a graphite die in order, and the die is subsequently heated by using a direct hot press or a spark plasma sintering method. Using this process, the sandwich structure composed of the electrode material, barrier material and thermoelectric material powder is thus formed at one time. The main advantages of this method include its simplicity and the resulting high interface bonding strength. Thermal spraying, a newer technology, is more efficient and scalable as compared to diffusion welding. As shown in Figure 1.13b, for the thermal spraying process, the thermoelectric material is cut to the desired dimensions for the thermoelectric legs and then plugged into a prefabricated frame made of a polymer or a ceramic. The metal electrode material and the barrier layer material can then be directly sprayed onto the contact surfaces of the thermoelectric legs. Thermal spraying is conducive to the simultaneous realization of interface processing and module integration for the device. Additionally, joins prepared by this method show higher stability at high temperatures (Zhang et al. 2016).

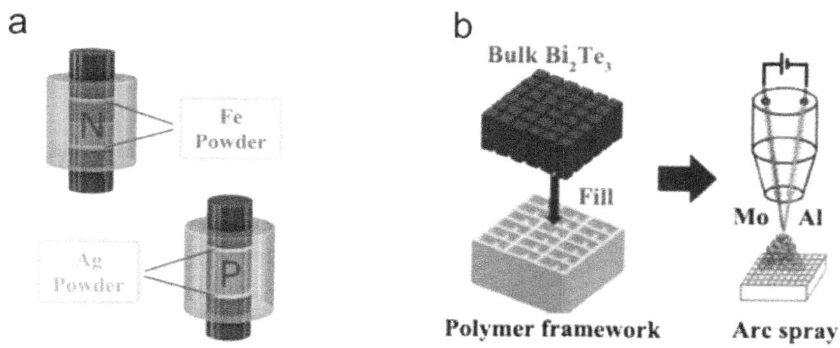

Figure 1.13. *Schematic diagrams of two major technologies for joining thermoelectric legs and metal contact layers: (a) diffusion welding. Reprinted from Ying et al. (2021) and (b) thermal spraying. Reprinted from Zhang et al. (2016). For a color version of this figure, see www.iste.co.uk/akinaga/thermoelectric2.zip*

After the metal electrode layer and the barrier layer are successfully connected to the thermoelectric legs, this working structure needs to be connected to the base (i.e. the substrate and the Cu conductive strips, which are generally joined in advance) by soldering or brazing. The selections of solder, interface pretreatment and soldering method are critical since the solder exhibits the highest CTE and the lowest yield strength among the interface components. The solder, in the form of a powder, paste or foil, is placed at the interface between the working structure and the base and should be heated and pressed to promote melting, and it is then cooled to firmly join the two parts (Kähler et al. 2014; Tewolde et al. 2015). Based on the above, one basic requirement for the solder is that its melting point should be lower than that of the thermoelectric legs and the conductive strips. Given that the maximum stress level is located on the thermoelectric legs near the hot-side plate, the use of a low-yield soldering alloy at the hot side and the promotion of plastic deformation in the solder can result in considerable thermal stress relief. The melting point of the solder should also be higher than the operating temperature to prevent interface softening or failure of the module during service. Additionally, soldering thickness has an impact on the temperature gradient and thermal stress distribution. An optimal soldering thickness will achieve both high conversion efficiency and low thermal stress (Wu et al. 2014). Currently, the most used solder materials are Sn-based alloys, such as Sn–Ag, Sn–Cu and Au–Sn.

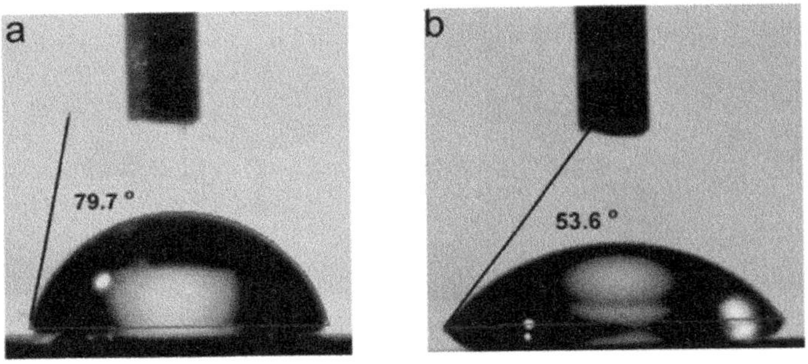

Figure 1.14. *Contact angle of a water drop on the surface of a nanostructured $Bi_2Te_{3-y}Se_y$ thermoelectric alloy (a) before and (b) after post-CMP (chemical mechanical polishing) cleaning (reprinted from Feng et al. (2011))*

It is worth noting that the quality of the soldering interface is not only associated with the type of solder and the soldering process but also with the cleanness of the interface. Careful surface cleaning is beneficial for removing surface oxides and hence reducing the contact resistance. In addition, surface cleaning can increase wettability significantly, which is beneficial for promoting welding. In a typical example, the contact angle of water dripped onto a Bi_2Te_3 surface was reduced from 79.7° to 53.6° (i.e. hydrophilicity increases) after a five-step treatment combining chemical and physical methods, as shown in Figure 1.14 (Feng et al. 2011). An interface incorporating solder has an inevitable shortcoming since it is easy to introduce contamination during the soldering process and form undesired phases which are harmful to the thermoelectric legs. In addition to applying a metal contact layer and a barrier layer, which can reduce the deterioration on the thermoelectric legs caused by solder to a certain extent, some new joining processes, such as non-soldering-based joining, have been developed and investigated (Malik et al. 2017). The solder composition could also be adjusted appropriately for different thermoelectric compositions.

1.10. Advanced thermoelectric module case studies

In the previous section, an overview of thermoelectric device design and the major related issues was provided. Generally, the design principles described above can be applied to most thermoelectric materials. However, for particular materials, targeted strategies are required to address issues such as thermal expansion, thermal stress, bonding strength, contact resistance and interfacial reaction or diffusion due to the differences in these materials' unique characteristics and specific operating temperature ranges. Hence, the reliability-related issues of some advanced thermoelectric modules and the corresponding strategies to address these issues are introduced in the following section.

1.10.1. Bi_2Te_3

Since they are among the most successful commercial thermoelectric materials, Bi_2Te_3-based materials have been studied extensively for decades. Usually, Ni is applied as a metal contact layer (3~10 μm in thickness) in Bi_2Te_3-based devices and the diffusion of Ni at the interface is negligible when the operating temperature is under 473 K. Additionally, Ni can effectively inhibit the formation of brittle Sn–Te intermetallic compounds resulting from the reaction between Sn-containing solder and the thermoelectric matrix. Therefore, as a metal contact layer, Ni is appropriate for refrigeration applications. However, when it is applied to thermoelectric power generators, a number of problems appear. Park et al. (2012) carried out a thermal cycling test from 303 K to 433 K on a commercial Bi_2Te_3 module, as shown in Figure 1.15a. During the first 2,000 cycles, the power output was found to remain consistent with the prediction value, but between 2,000 and 6,000 cycles, there were 11% and 12% reductions in the power output for load resistance values of 47 Ω and 1,000 Ω, respectively, as shown in Figure 1.15b.

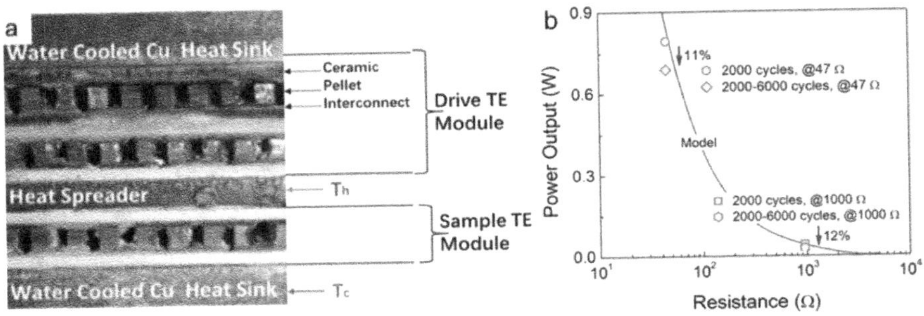

Figure 1.15. *(a) Thermal cycling test system configuration for a commercial Bi_2Te_3 module. Reprinted from Park et al. (2012). (b) Power output as a function of load resistance. Two power output data point (maximum and minimum values) for each load resistance value indicate the degradation of the thermoelectric module during thermal cycling. Data obtained from Park et al. (2012). For a color version of this figure, see www.iste.co.uk/akinaga/thermoelectric2.zip*

Liu et al. (2013) studied the interface reaction between Ni and n-type $Bi_2Te_{2.7}Se_{0.3}$ and that between Ni and p-type $Bi_{0.4}Sb_{1.6}Te_3$ by analyzing the interface microstructure of each sample hot-pressed at different temperatures, as shown in Figure 1.16. It is clear that the thickness of the interface reaction layer increases with increasing hot-pressing temperature. The thick diffusion layer is beneficial for strong bonding at the interface, but the diffusion of Ni was found to form a p-type region within the n-type $Bi_2Te_{2.7}Se_{0.3}$ leg, resulting in a large contact resistance for the n-type interface (~210 $\mu\Omega$ cm^2), while the contact resistance remains low for the p-type $Bi_{0.4}Sb_{1.6}Te_3$ leg (<1 $\mu\Omega$ cm^2) (Liu et al. 2013). Applying a barrier layer between the Ni contact layer and these thermoelectric materials is advised to address severe interface reactions at high temperatures. Additionally, seeking out new materials for the metal contact layer could be an alternative for this interface design. Research on devices based on Bi_2Te_3 remains ongoing and is currently mainly focused on improving output performance. For instance, Nozariasbmarz et al. (2020) reported that a record 8% conversion efficiency and 2.1 W/cm^2 power density were obtained in an n-/p-type Bi_2Te_3-based module for waste heat recovery (298 K~523 K). In addition, a Bi_2Te_3/half-Heusler-based segmented unicouple and a Bi_2Te_3/skutterudite segmented module also show remarkable energy conversion efficiency around 12% (Li et al. 2020d). Unfortunately, results from aging tests and long-term service tests were not provided in these reports.

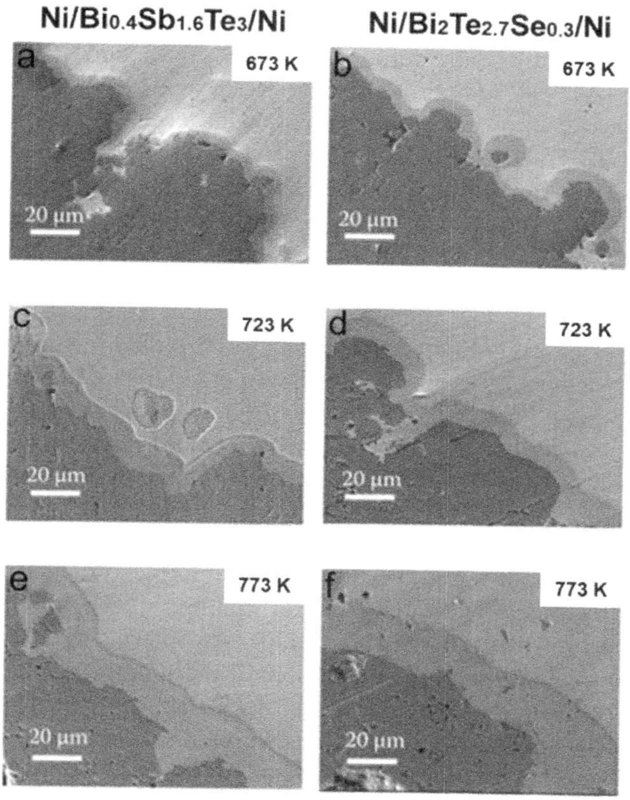

Figure 1.16. *Microstructures of contact interfaces made by directly hot-pressing Ni powder together with p-type $Bi_{0.4}Sb_{1.6}Te_3$ or n-type $Bi_2Te_{2.7}Se_{0.3}$ thermoelectric powder. Ni/$Bi_{0.4}Sb_{1.6}Te_3$/Ni hot-pressed at (a) 673 K, (c) 723 K and (e) 773 K. Ni/$Bi_2Te_{2.7}Se_{0.3}$/Ni hot-pressed at (b) 673 K, (d) 723 K and (f) 773 K. In each image, the dark region is nickel and the bright region is the thermoelectric material. Reprinted from Liu et al. (2013)*

1.10.2. *$Mg_3(Sb, Bi)_2$*

N-type $Mg_3(Sb, Bi)_2$-based materials have attracted tremendous attention in recent years because of their promising thermoelectric performance, low cost and good mechanical properties. It is worth mentioning that $Mg_3(Sb, Bi)_2$-based devices can be used for different applications in several temperature ranges, for example, room-temperature thermoelectric cooling (Mao et al. 2019a; Shu et al. 2019a), low-grade heat recovery (Imasato et al. 2019b; Liang et al. 2021) and power generation in the medium-temperature range (Xu et al. 2021), by tuning the ratio of Sb and Bi to either Sb- or Bi-rich. As an emerging material system, $Mg_3(Sb, Bi)_2$-based thermoelectrics have received little attention in terms of interface research. Zhu et al. (2019) were the first to report on the $Mg_3(Sb, Bi)_2$ contact interface in 2019. In their study, a good junction interface with low contact resistivity and strong bonding strength was formed between $Mg_{3.1}Co_{0.1}Sb_{1.5}Bi_{0.49}Te_{0.01}$ and Fe using a direct hot-pressing method, which was found to lead to a high energy conversion efficiency of 10.6% between 373 K and 773 K. Fe has since been successfully used as a metal contact electrode layer in a thermoelectric cooling system (Mao et al. 2019a) and a low-grade waste heat recovery device (Liang et al. 2021). In a later study, Ni was found to play a role similar to that of Fe as a contact layer (Mao et al. 2019a). Ni can also ensure a low electrical contact resistance for soldering once it is electroplated on both sides of the $Mg_3(Sb, Bi)_2$ thermoelectric leg (Shang et al. 2020b; Xu et al. 2021). On the other hand, the Fe contact layer could degrade upon long-term aging, which may be attributed to the mismatched CTE values and the undesirable interface reaction between the thermoelectric material and Fe. Thus, 304 stainless steel was introduced as a potential replacement by Yin et al. (2020). An interface constructed using 304 stainless steel and a $Mg_3(Sb, Bi)_2$-based thermoelectric material shows smaller changes in its electrical and mechanical properties with aging compared to that constructed using Fe due to its suppressed diffusion and enhanced expansion coefficient match, as shown in Figure 1.17. A high conversion efficiency of up to 9% at a temperature difference of 370 K in a single leg incorporating this $Mg_3(Sb, Bi)_2$-based thermoelectric material and 304 stainless steel was reported in the same study. Unfortunately, no studies of the long-term operation of an integrated module based on $Mg_3(Sb, Bi)_2$ have yet been reported. In short, despite the good thermoelectric properties of $Mg_3(Sb, Bi)_2$-based materials, only with abundant contact layer research and module evaluation can these compounds achieve a large-scale practical application.

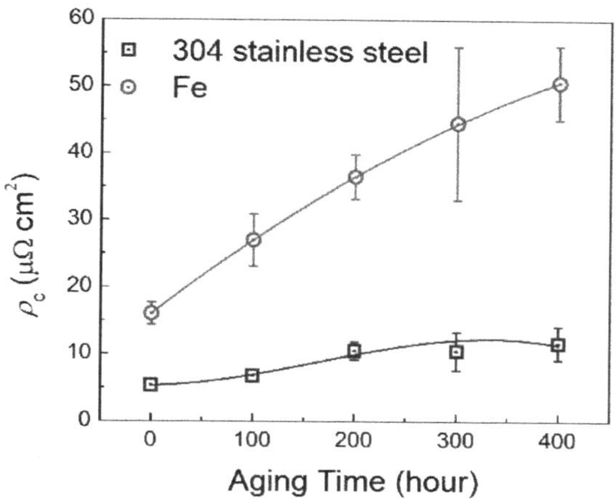

Figure 1.17. *Aging-time-dependent contact resistivity of $Mg_{3.2}Sb_{1.5}Bi_{0.49}Te_{0.01}$/304 stainless steel and $Mg_{3.2}Sb_{1.5}Bi_{0.49}Te_{0.01}$/Fe junctions aged at 523 K. Data obtained from Yin et al. (2020). For a color version of this figure, see www.iste.co.uk/akinaga/thermoelectric2.zip*

1.10.3. *GeTe*

Research on GeTe as a thermoelectric material began as early as the 1960s, but it was not studied in depth for long since it has an excessively high carrier concentration (~10^{21} cm^{-3}) and an undesired phase change at ~700 K. In recent years, GeTe has returned as a focus of thermoelectric research because it was found that its carrier concentration can be reduced to an ideal range (~10^{20} cm^{-3}) by doping with Sb or Bi (Liu et al. 2018; Zheng et al. 2018). Additionally, its undesired phase transition can be suppressed by element doping, as discussed in section 1.5. GeTe has subsequently become a promising candidate for use in thermoelectric generators for waste heat recovery in the mid-temperature range. Although the thermoelectric performance of GeTe has been greatly improved, device research is currently lagging significantly, delaying actual applications. At high temperatures, fast chemical reactions and diffusion can occur between the metal electrodes and the thermoelectric legs in GeTe-based thermoelectric modules. Thus, a barrier layer is required. An Al–Si alloy was recently proposed as a diffusion barrier for GeTe-based thermoelectric legs by Li et al. (2020b). In their study, a good connection between $Al_{66}Si_{34}$ and $Ge_{0.9}Sb_{0.1}TeB_{0.01}$ was obtained using the spark plasma sintering method and the resulting thermoelectric legs underwent aging testing for 16 days at 773 K. After the

16-day test, their contact resistivity increased slightly from 20.7 to 26.1 μΩ·cm^2. The thickness of the interface diffusion layer increased quickly from ~8 to ~18 μm during the first 4 days of aging but slowed down after 8 days of aging. Overall, the thickness of the interface diffusion layer increased to ~23 μm after 16 days of aging, as shown in Figure 1.18a–c. The shear strength of the thermoelectric legs also increased from 26.6 MPa to 41.7 MPa after the aging test, which may be attributed to the thicker diffusion layer. Xing et al. (2021) chose Mo as an effective diffusion barrier material for GeTe-based thermoelectric legs by testing 12 pure metals (Cr, Hf, Nb, Ti, Mo, Ni, Ta, Zr, Al, Co, V and Fe). The result of resistance line scanning across the interface of GeTe/Mo/GeTe indicates that an extraordinarily low value (less than 1 μΩ·cm^2) can be obtained. Furthermore, a $Ge_{0.92}Sb_{0.04}Bi_{0.04}Te_{0.95}Se_{0.05}$/Mo/Ni thermoelectric leg was constructed and underwent aging testing at 800 K, and its interface microstructure was found to remain stable after 10 days of testing, as shown in Figure 1.18d–f. Finally, an eight-couple $Ge_{0.92}Sb_{0.04}Bi_{0.04}Te_{0.95}Se_{0.05}/Yb_{0.3}Co_4Sb_{12}$ thermoelectric module was fabricated and achieved an energy conversion efficiency of 7.8% and a power density of 1.1 W·cm^{-2} under a temperature difference of 500 K.

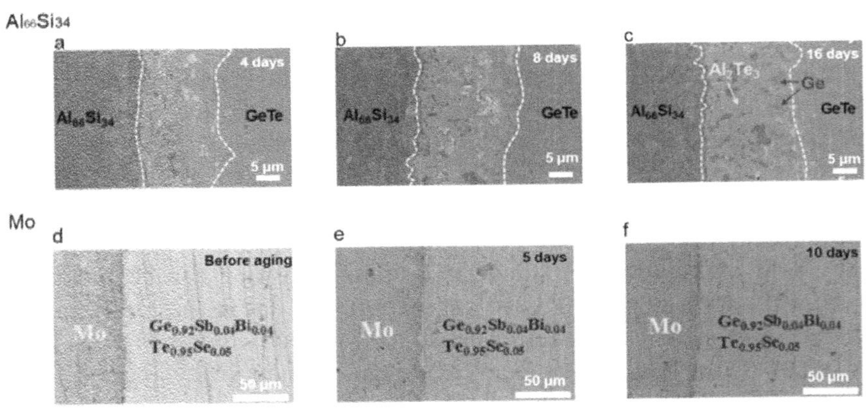

Figure 1.18. *Scanning electron microscopy images of the interface area of $Ge_{0.9}Sb_{0.1}TeB_{0.01}/Al_{66}Si_{34}$ thermoelectric legs after (a) 4 days, (b) 8 days and (c) 16 days of aging at 773 K. Reprinted from Li et al. (2020b). Scanning electron microscopy images of the interface area of $Ge_{0.92}Sb_{0.04}Bi_{0.04}Te_{0.95}Se_{0.05}$/Mo/Ni thermoelectric legs (d) before aging and after aging (e) 5 days and (f) 10 days at 800 K. Reprinted from Xing et al. (2021). For a color version of this figure, see www.iste.co.uk/akinaga/thermoelectric2.zip*

1.10.4. *Skutterudites*

Filled skutterudites show promising thermoelectric and mechanical properties for mid-temperature power generation. The contact design of skutterudites has been extensively examined in recent decades. Pure metals and alloys like Ni, Ti, Zr, Nb, Mo-Cu and Co-Fe-Ni have been employed and evaluated as metal contact layers (Zhang 2018; Liu and Bai 2019). Although Ni is a good candidate for most thermoelectric materials in interfacial design, it is not suitable for skutterudites due to its larger CTE. Therefore, a buffer layer is needed when Ni is applied as the metal contact layer for skutterudites. Ti- and Zr-based contact layers often suffer from a severe interfacial chemical reaction during the aging process, so they do not provide adequate stability for long-term power generation service. The Mo–Cu alloy is another potential candidate for the metal contact layer because of its high electrical and thermal conductivity. Additionally, its composition can be easily tuned to match the CTE of the filled skutterudite (Zhang 2018). However, it is very difficult to bond the Mo–Cu alloy onto the surface of a skutterudite thermoelectric leg. In fact, the most significant challenge in choosing a metal contact layer for skutterudites is that their main element, Sb, is highly reactive, which may cause a severe interfacial reaction and diffusion at high temperatures, and thus significantly degrade the device performance. To balance effective bonding and a durable interface, Chu et al. (2020) recently proposed a criterion to predict the bonding behavior as well as the interfacial reliability of $CoSb_3$-based filled skutterudite/metal electrode interfaces by combining the interfacial reaction energy (E_{IR}) value of a metal with its Sb migration activation energy barrier (E_{Mig}) value, as shown in Figure 1.19a. Generally, E_{IR} should be negative to promote the progress of the interface reaction and produce a mechanical force at the interface, while E_{Mig} should be large enough to limit Sb diffusion and prevent the interface reaction layer from growing too quickly. Based on their simulation results, they defined a "sweet spot" area as shown in Figure 1.19a, and metals in this region were determined to be potential candidates for the metal contact layer. To validate the proposed criterion, Nb was chosen as the metal contact layer to fabricate a $CoSb_3$-based filled skutterudite thermoelectric leg. Their experimental results showed robust bonding with no obvious diffusion under thermal cycles, confirming that the interface reaction was inhibited and the interfacial resistivity was reduced. The derived module also showed excellent stability in output power as well as internal resistance after 846 hours of service testing, as shown in Figure 1.19b. In addition, work–function matching is another effective guiding principle for interfacial design that was proposed by Jie et al. (2015). They reported that the matching of the work function between a metal contact layer and a filled skutterudite can play a crucial role in determining the electrical contact resistance based on their study of the interface behavior between the p-type $Ce_{0.45}Nd_{0.45}Fe_{3.5}Co_{0.5}Sb_{12}$ and a Cr–Fe–Co alloy or a Cr–Fe–Ni alloy.

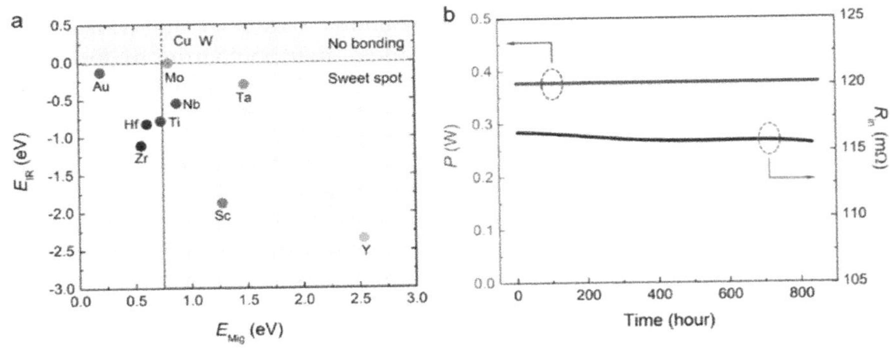

Figure 1.19. *(a) Correlation map of interfacial reaction energy (E_{IR}) and Sb migration activation energy barrier (E_{Mig}) values of different metals. (b) Output power and internal resistance of an eight-pair $CoSb_3$-based filled skutterudite module fabricated using Nb as the barrier layer under long-term service conditions with T_h = 818 K and T_c = 308 K. Data obtained from Chu et al. (2020). For a color version of this figure, see www.iste.co.uk/akinaga/thermoelectric2.zip*

1.11. Summary and outlook

In comparison with zT improvement among various thermoelectric materials over the past decades, the development process from thermoelectric materials to devices has been much slower. For the materials' optimization, more effort should be devoted to improving the thermal stability and mechanical robustness of the materials with relatively high zT because commercialization will not be possible until such improvement is achieved. To accelerate commercialization, a multi-criteria evaluation method needs to be developed based on the performance and reliability requirements. Since the interfacial layer plays an important role in performance maximization and structure integrity, it is meaningful to construct a rational model based on experiment and simulation to guide the design of high-performance contact layers. For device development, evaluating the performance of each lab-prepared module based on multiple industrial standards (e.g. vibration testing, mechanical shock testing, temperature cycle endurance testing, etc.) can help to determine commercial viability in realizing actual applications. Most importantly, it is time to try all possible device applications to demonstrate some of the promise of the past decades of materials research to build a case for economical adaption of this technology to real applications. With enough practical applications of thermoelectric devices, more research will be funded and can be carried out, which in turn will enhance the device applications. Such a healthy, self-propelling cycle is what the thermoelectric research community needs going forward.

1.12. References

Al-Merbati, A.S., Yilbas, B.S., Sahin, A.Z. (2013). Thermodynamics and thermal stress analysis of thermoelectric power generator: Influence of pin geometry on device performance. *Appl. Therm. Eng.*, 50(1), 683–692.

Barako, M.T., Park, W., Marconnet, A.M., Asheghi, M., Goodson, K.E. (2012). Thermal cycling, mechanical degradation, and the effective figure of merit of a thermoelectric module. *J. Electron. Mater.*, 42(3), 372–381.

Bohra, A.K., Bhatt, R., Bhattacharya, S., Basu, R., Ahmad, S., Singh, A., Aswal, D.K., Gupta, S.K. (2016). Study of thermal stability of Cu_2Se thermoelectric material. *AIP Conf. Proc.*, 1713, 110010.

Bohra, A.K., Bhatt, R., Singh, A., Bhattacharya, S., Basu, R., Bhatt, P., Navaneethan, M., Sarkar, S.K., Anwar, S., Muthe, K.P. et al. (2020). Stabilizing thermoelectric figure-of-merit of superionic conductor Cu_2Se through nanoinclusions. *Phys. Status Solidi (RRL)*, 14, 2000102.

Chen, L.D., Dong, H.L., Li, X.Y., Huang, X.Y., Jiang, W., Tang, Y.S., Xia, X.G., Liao, J.C., He, L. (2013). Skutterudite based thermoelectric material, thermal protection coating for devices and preparation method thereof. Patent, CN103146301A.

Chowdhury, I., Prasher, R., Lofgreen, K., Chrysler, G., Narasimhan, S., Mahajan, R., Koester, D., Alley, R., Venkatasubramanian, R. (2009). On-chip cooling by superlattice-based thin-film thermoelectrics. *Nat. Nanotechnol.*, 4(4), 235–238.

Chu, J., Huang, J., Liu, R., Liao, J., Xia, X., Zhang, Q., Wang, C., Gu, M., Bai, S., Shi, X. et al. (2020). Electrode interface optimization advances conversion efficiency and stability of thermoelectric devices. *Nat. Commun.*, 11(1), 2723.

Dasgupta, T., Stiewe, C., Sesselmann, A., Yin, H., Iversen, B.B., Mueller, E. (2013). Thermoelectric studies in β-Zn_4Sb_3-the complex interdependence between thermal stability, thermoelectric transport, and zinc content. *J. Appl. Phys.*, 113, 103708.

Deng, S., Chen, Z., Li, D., Liu, H., Tang, Y., Shen, L., Deng, S. (2017). Effect of Pb doped on thermal stability and electrical transport properties of single crystalline β-Zn_4Sb_3. *Physica B*, 526, 155–159.

Dong, H., Li, X., Tang, Y., Zou, J., Huang, X., Zhou, Y., Jiang, W., Zhang, G.-J., Chen, L. (2012). Fabrication and thermal aging behavior of skutterudites with silica-based composite protective coatings. *J. Alloys Compd.*, 527, 247–251.

El-Genk, M.S., Saber, H.H., Caillat, T., Sakamoto, J. (2006). Tests results and performance comparisons of coated and un-coated skutterudite based segmented unicouples. *Energy Convers. Manag.*, 47(2), 174–200.

Erturun, U., Erermis, K., Mossi, K. (2015). Influence of leg sizing and spacing on power generation and thermal stresses of thermoelectric devices. *Appl. Energy*, 159, 19–27.

Fabián-Mijangos, A., Min, G., Alvarez-Quintana, J. (2017). Enhanced performance thermoelectric module having asymmetrical legs. *Energy Convers. Manag.*, 148, 1372–1381.

Feng, H.-P., Yu, B., Chen, S., Collins, K., He, C., Ren, Z.F., Chen, G. (2011). Studies on surface preparation and smoothness of nanostructured Bi_2Te_3-based alloys by electrochemical and mechanical methods. *Electrochim. Acta*, 56(8), 3079–3084.

Feng, S.P., Chang, Y.H., Yang, J., Poudel, B., Yu, B., Ren, Z., Chen, G. (2013). Reliable contact fabrication on nanostructured Bi_2Te_3-based thermoelectric materials. *Phys. Chem. Chem. Phys.*, 15(18), 6757–6762.

Hao, Y., Pedersen, B.L., Iversen, B.B. (2011). Thermal stability of high performance thermoelectric β-Zn_4Sb_3 in argon. *Eur. J. Inorg. Chem.*, 17, 2733–2737.

He, R., Schierning, G., Nielsch, K. (2018). Thermoelectric devices: A review of devices, architectures, and contact optimization. *Adv. Mater. Technol.*, 3(4), 1700256.

Hong, M., Zheng, K., Lyv, W., Li, M., Qu, X., Sun, Q., Xu, S., Zou, J., Chen, Z.-G. (2020). Computer-aided design of high-efficiency GeTe-based thermoelectric devices. *Energy Environ. Sci.*, 13(6), 1856–1864.

Imasato, K., Wood, M., Kuo, J.J., Snyder, G.J. (2018). Improved stability and high thermoelectric performance through cation site doping in n-type La-doped $Mg_3Sb_{1.5}Bi_{0.5}$. *J. Mater. Chem. A*, 6(41), 19941–19946.

Imasato, K., Kang, S.D., Snyder, G.J. (2019). Exceptional thermoelectric performance in $Mg_3Sb_{0.6}Bi_{1.4}$ for low-grade waste heat recovery. *Energy Environ. Sci.*, 12, 965–971.

Ioffe, A.F. (1957). *Semiconductor Thermoelements and Thermoelectric Cooling*. Infosearch Ltd, London.

Jia, X. and Gao, Y. (2014). Estimation of thermoelectric and mechanical performances of segmented thermoelectric generators under optimal operating conditions. *Appl. Therm. Eng.*, 73(1), 335–342.

Jie, Q., Ren, Z., Chen, G. (2015). Fabrication of stable electrode/diffusion barrier layers for thermoelectric filled skutterudite devices. Patent, US9722164B2.

Jørgensen, L.R., Zhang, J., Zeuthen, C.B., Iversen, B.B. (2018). Thermal stability of $Mg_3Sb_{1.475}Bi_{0.475}Te_{0.05}$ high performance n-type thermoelectric investigated through powder X-ray diffraction and pair distribution function analysis. *J. Mater. Chem. A*, 6(35), 17171–17176.

Joshi, G. and Poudel, B. (2016). Efficient and robust thermoelectric power generation device using hot-pressed metal contacts on nanostructured half-Heusler alloys. *J. Electron. Mater.*, 45(12), 6047–6051.

Kähler, J., Stranz, A., Waag, A., Peiner, E. (2014). Thermoelectric coolers with sintered silver interconnects. *J. Electron. Mater.*, 43(6), 2397–2404.

Kambe, M., Jinushi, T., Ishijima, Z. (2010). Encapsulated thermoelectric modules and compliant pads for advanced thermoelectric systems. *J. Electron. Mater.*, 39(9), 1418–1421.

Kang, S.D., Pohls, J.-H., Aydemir, U., Qiu, P., Stoumpos, C.C., Hanus, R., White, M.A., Shi, X., Chen, L., Kanatzidis, M.G. et al. (2017). Enhanced stability and thermoelectric figure-of-merit in copper selenide by lithium doping. *Mater. Today Phys.*, 1, 7–13.

Karri, N.K. and Mo, C. (2018). Structural reliability evaluation of thermoelectric generator modules: Influence of end conditions, leg geometry, metallization, and processing temperatures. *J. Electron. Mater.*, 47(10), 6101–6120.

Kim, C.N. (2018). Development of a numerical method for the performance analysis of thermoelectric generators with thermal and electric contact resistance. *Appl. Therm. Eng.*, 130, 408–417.

Kim, H.S., Wang, T., Liu, W., Ren, Z. (2016). Engineering thermal conductivity for balancing between reliability and performance of bulk thermoelectric generators. *Adv. Funct. Mater.*, 26(21), 3678–3686.

Kim, H.S., Liu, W., Ren, Z. (2017). The bridge between the materials and devices of thermoelectric power generators. *Energy Environ. Sci.*, 10(1), 69–85.

Kuo, J.J., Yu, Y., Kang, S.D., Cojocaru-Mirédin, O., Wuttig, M., Snyder, G.J. (2019). Mg deficiency in grain boundaries of *n*-type Mg_3Sb_2 identified by atom probe tomography. *Adv. Mater. Inter.*, 6(13), 1900429.

Li, F., Wei, T.-R., Kang, F., Li, J.-F. (2014). Thermal stability and oxidation resistance of BiCuSeO based thermoelectric ceramics. *J. Alloys Compd.*, 614, 394–400.

Li, J., Jia, F., Zhang, S., Zheng, S., Wang, B., Chen, L., Lu, G., Wu, L. (2019). The manipulation of substitutional defects for realizing high thermoelectric performance in Mg_3Sb_2-based Zintl compounds. *J. Mater. Chem. A*, 7(33), 19316–19323.

Li, J., Zhang, S., Jia, F., Zheng, S., Shi, X., Jiang, D., Wang, S., Lu, G., Wu, L., Chen, Z.-G. (2020a). Point defect engineering and machinability in n-type Mg_3Sb_2-based materials. *Mater. Today Phys.*, 15, 100269.

Li, J., Zhao, S., Chen, J., Han, C., Hu, L., Liu, F., Ao, W., Li, Y., Xie, H., Zhang, C. (2020b). Al–Si alloy as a diffusion barrier for GeTe-based thermoelectric legs with high interfacial reliability and mechanical strength. *ACS Appl. Mater. Interfaces*, 12(16), 18562–18569.

Li, M., Hong, M., Tang, X., Sun, Q., Lyu, W.-Y., Xu, S.-D., Kou, L.-Z., Dargusch, M., Zou, J., Chen, Z.-G. (2020c). Crystal symmetry induced structure and bonding manipulation boosting thermoelectric performance of GeTe. *Nano Energy*, 73, 104740.

Li, W., Poudel, B., Nozariasbmarz, A., Sriramdas, R., Zhu, H., Kang, H.B., Priya, S. (2020d). Bismuth telluride/half-Heusler segmented thermoelectric unicouple modules provide 12% conversion efficiency. *Adv. Energy Mater.*, 10(38), 2001924.

Li, M., Sun, Q., Xu, S.D., Hong, M., Lyu, W.Y., Liu, J.X., Wang, Y., Dargusch, M., Zou, J., Chen, Z.G. (2021). Optimizing electronic quality factor toward high-performance $Ge_{1-x-y}Ta_xSb_yTe$ thermoelectrics: The role of transition metal doping. *Adv. Mater.*, 33(40), 2102575.

Liang, Z., Xu, C., Shang, H., Zhu, Q., Ding, F., Mao, J., Ren, Z. (2021). High thermoelectric energy conversion efficiency of a unicouple of n-type Mg_3Bi_2 and p-type Bi_2Te_3. *Mater. Today Phys.*, 19, 100413.

Lin, W.-C., Li, Y.-S., Wu, A.T. (2017). Study of diffusion barrier for solder/n-type Bi_2Te_3 and bonding strength for p- and n-type thermoelectric modules. *J. Electron. Mater.*, 47(1), 148–154.

Liu, W. and Bai, S. (2019). Thermoelectric interface materials: A perspective to the challenge of thermoelectric power generation module. *J. Mater.*, 5(3), 321–336.

Liu, H., Shi, X., Xu, F., Zhang, L., Zhang, W., Chen, L., Li, Q., Uher, C., Day, T., Snyder, G.J. (2012). Copper ion liquid-like thermoelectrics. *Nat. Mater.*, 11(5), 422–425.

Liu, W., Wang, H., Wang, L., Wang, X., Joshi, G., Chen, G., Ren, Z. (2013). Understanding of the contact of nanostructured thermoelectric n-type $Bi_2Te_{2.7}Se_{0.3}$ legs for power generation applications. *J. Mater. Chem. A*, 1, 13093–13100.

Liu, W., Jie, Q., Kim, H.S., Ren, Z. (2015). Current progress and future challenges in thermoelectric power generation: From materials to devices. *Acta Mater.*, 87, 357–376.

Liu, Z., Sun, J., Mao, J., Zhu, H., Ren, W., Zhou, J., Wang, Z., Singh, D.J., Sui, J., Chu, C.W. et al. (2018). Phase-transition temperature suppression to achieve cubic GeTe and high thermoelectric performance by Bi and Mn codoping. *Proc. Natl. Acad. Sci. USA*, 115(21), 5332–5337.

Liu, W., Shen, L., Shai, X., Tang, Y., Chen, Z., Sun, L., Ge, W., Deng, S. (2019). Effect of hole modulation on the electrical transport properties and thermal stability of single crystal α-Cu_2Se prepared by Bi-flux method. *J. Alloy. Compd.*, 791, 60–67.

Liu, Z.-Y., Zhu, J.-L., Tong, X., Niu, S., Zhao, W.-Y. (2020). A review of $CoSb_3$-based skutterudite thermoelectric materials. *J. Adv. Ceram.*, 9(6), 647–673.

Lu, J., Li, D., Liu, W., Shen, L., Chen, J., Ge, W., Deng, S. (2020a). Thermal stability and thermoelectric properties of Cd-doped nano-layered Cu_2Se prepared using NaCl flux method. *Chin. Phys. B*, 29, 127403.

Lu, R., Olvera, A., Bailey, T.P., Uher, C., Poudeu, F.P. (2020b). Nanoscale engineering of polymorphism in Cu_2Se-based composites. *ACS Appl. Mater. Interfaces*, 12(28), 31601–31611.

Malik, S.A., Hung, L.T., Van Nong, N. (2017). Solder free joining as a highly effective method for making contact between thermoelectric materials and metallic electrodes. *Mater. Today Energy*, 5, 305–311.

Mao, J., Shuai, J., Song, S., Wu, Y., Dallye, R., Zhou, J., Liu, Z., Sun, J., Zhang, Q., Cruz, C.D. et al. (2017a). Manipulation of ionized impurity scattering for achieving high thermoelectric performance in n-type Mg_3Sb_2-based materials. *Proc. Natl. Acad. Sci. USA*, 114(40), 10548–10553.

Mao, J., Wu, Y., Song, S., Zhu, Q., Shuai, J., Liu, Z., Pei, Y., Ren, Z. (2017b). Defect engineering for realizing high thermoelectric performance in n-type Mg_3Sb_2-based materials. *ACS Energy Lett.*, 2(10), 2245–2250.

Mao, J., Liu, Z., Zhou, J., Zhu, H., Zhang, Q., Chen, G., Ren, Z. (2018). Advances in thermoelectrics. *Adv. Phys.*, 67(2), 69–147.

Mao, J., Zhu, H., Ding, Z., Liu, Z., Gamage, G.A., Chen, G., Ren, Z. (2019a). High thermoelectric cooling performance of n-type Mg_3Bi_2-based materials. *Science*, 365(6452), 495–498.

Mao, T., Qiu, P., Du, X., Hu, P., Zhao, K., Xiao, J., Shi, X., Chen, L. (2019b). Enhanced thermoelectric performance and service stability of Cu_2Se via tailoring chemical compositions at multiple atomic positions. *Adv. Funct. Mater.*, 30(6), 1908315.

Mejri, M., Malard, B., Thimont, Y., Romanjek, K., Ihou Mouko, H., Estournès, C. (2020). Thermal stability of $Mg_2Si_{0.55}Sn_{0.45}$ for thermoelectric applications. *J. Alloys Compd.*, 846, 156413.

Moghaddam, A.O., Shokuhfar, A., Zhang, Y., Zhang, T., Cadavid, D., Arbiol, J., Cabot, A. (2019). Ge-doped ZnSb/β-Zn_4Sb_3 nanocomposites with high thermoelectric performance. *Adv. Mater. Interfaces*, 6(18), 1900467.

Music, D., Geyer, R.W., Keuter, P. (2016). Thermomechanical response of thermoelectrics. *Appl. Phys. Lett.*, 109(22), 223903.

Ni, J.E., Case, E.D., Schmidt, R.D., Wu, C.-I., Hogan, T.P., Trejo, R.M., Kirkham, M.J., Lara-Curzio, E., Kanatzidis, M.G. (2013). The thermal expansion coefficient as a key design parameter for thermoelectric materials and its relationship to processing-dependent bloating. *J. Mater. Sci.*, 48(18), 6233–6244.

Nozariasbmarz, A., Poudel, B., Li, W., Kang, H.B., Zhu, H., Priya, S. (2020). Bismuth telluride thermoelectrics with 8% module efficiency for waste heat recovery application. *Science*, 23(7), 101340.

Nunn, R., Qiu, P., Yin, M., Chen, H., Hanus, R., Song, Q., Zhang, T., Chou, M.-Y., Agne, M.T., He, J. et al. (2017). Ultrahigh thermoelectric performance in Cu_2Se-based hybrid materials with highly dispersed molecular CNTs. *Energy Environ. Sci.*, 10, 1928–1935.

Park, W., Marconnet, A.M., Asheghi, M., Goodson, K.E. (2012). *13th InterSociety Conference on Thermal and Thermomechanical Phenomena in Electronic Systems*. IEEE, San Diego.

Rausch, E., Balke, B., Ouardi, S., Felser, C. (2016). Long-term stability of (Ti/Zr/Hf)$CoSb_{1-x}Sn_x$ thermoelectric p-type half-Heusler compounds upon thermal cycling. *Energy Technol.*, 3, 1217–1224.

Ravi, V., Firdosy, S., Caillat, T., Brandon, E., Van Der Walde, K., Maricic, L., Sayir, A. (2009). Thermal expansion studies of selected high-temperature thermoelectric materials. *J. Electron. Mater.*, 38(7), 1433–1442.

Ren, Z., Lan, Y., Zhang, Q. (2018). *Advanced Thermoelectrics: Materials, Contacts, Devices, and Systems*. CRC Press, New York.

Rogl, G. and Rogl, P. (2017). How nanoparticles can change the figure of merit, ZT, and mechanical properties of skutterudites. *Mater. Today Phys.*, 3, 48–69.

Rogl, G., Zhang, L., Rogl, P., Grytsiv, A., Falmbigl, M., Rajs, D., Kriegisch, M., Müller, H., Bauer, E., Koppensteiner, J. et al. (2010). Thermal expansion of skutterudites. *Int. J. Appl. Phys.*, 107(4), 043507.

Rosi, F.D. and Bernoff, R.A. (1962). Method and materials for obtaining low resistance bonds to thermoelectric bodies. Patent, US3037064A.

Rowe, D.M. (1995). *Thermoelectrics Handbook*. CRC Press, London.

Sakamoto, J.S., Snyder, G.J., Calliat, T., Fleurial, J.P., Jones, S.M., Palk, J.A. (2008). System and method for suppressing sublimation using opacified aerogel. Patent, US7461512B2.

Sales, B.C., Mandrus, D., Williams, R.K. (1996). Filled skutterudite antimonides: A new class of thermoelectric materials. *Science*, 272(5266), 1325–1328.

Sarhadi, A., Bjørk, R., Pryds, N. (2015). Optimization of the mechanical and electrical performance of a thermoelectric module. *J. Electron. Mater.*, 44(11), 4465–4472.

Shai, X., Deng, S., Shen, L., Meng, D., Li, D., Zhang, Y., Jiang, X. (2015). Preparation, thermal stability, and electrical transport properties of In/Sn codoped b-Zn_4Sb_3 single crystal. *Phys. Status Solidi B*, 252(4), 1–5.

Shang, H., Liang, Z., Xu, C., Mao, J., Gu, H., Ding, F., Ren, Z. (2020a). N-type $Mg_3Sb_{2-x}Bi_x$ alloys as promising thermoelectric materials. *Research*, 1219461.

Shang, H., Liang, Z., Xu, C., Song, S., Huang, D., Gu, H., Mao, J., Ren, Z., Ding, F. (2020b). N-type $Mg_3Sb_{2-x}Bi_x$ with improved thermal stability for thermoelectric power generation. *Acta Mater.*, 201, 572–579.

Shi, X., Sun, C., Bu, Z., Zhang, X., Wu, Y., Lin, S., Li, W., Faghaninia, A., Jain, A., Pei, Y. (2019a). Revelation of inherently high mobility enables Mg_3Sb_2 as a sustainable alternative to n-Bi_2Te_3 thermoelectrics. *Adv. Sci.*, 6(16), 1802286.

Shi, X., Sun, C., Zhang, X., Chen, Z., Lin, S., Li, W., Pei, Y. (2019b). Efficient Sc-doped $Mg_{3.05-x}Sc_xSbBi$ thermoelectrics near room temperature. *Chem. Mater.*, 31(21), 8987–8994.

Shi, X., Zhao, T., Zhang, X., Sun, C., Chen, Z., Lin, S., Li, W., Gu, H., Pei, Y. (2019c). Extraordinary n-type Mg_3SbBi thermoelectrics enabled by yttrium doping. *Adv. Mater.*, 31(36), 1903387.

Shi, D., Geng, Z., Shi, L., Li, Y., Lam, K.H. (2020). Thermal stability study of $Cu_{1.97}Se$ superionic thermoelectric material. *J. Mater. Chem. C*, 8, 10221–10228.

Shittu, S., Li, G., Zhao, X., Ma, X. (2020). Review of thermoelectric geometry and structure optimization for performance enhancement. *Appl. Energy*, 268(150), 115075.

Shu, R., Zhou, Y., Wang, Q., Han, Z., Zhu, Y., Liu, Y., Chen, Y., Gu, M., Xu, W., Wang, Y. et al. (2019). $Mg_{3+\delta}Sb_xBi_{2-x}$ family: A promising substitute for the state-of-the-srt n-type thermoelectric materials near room temperature. *Adv. Funct. Mater.*, 29(4), 1807235.

Shuai, J., Mao, J., Song, S., Zhang, Q., Chen, G., Ren, Z. (2017a). Recent progress and future challenges on thermoelectric Zintl materials. *Mater. Today Phys.*, 1, 74–95.

Shuai, J., Mao, J., Song, S., Zhu, Q., Sun, J., Wang, Y., He, R., Zhou, J., Chen, G., Singh, D.J. et al. (2017b). Tuning the carrier scattering mechanism to effectively improve the thermoelectric properties. *Energy Environ. Sci.*, 10(3), 799–807.

Shuai, J., Ge, B., Mao, J., Song, S., Wang, Y., Ren, Z. (2018). Significant role of Mg stoichiometry in designing high thermoelectric performance for $Mg_3(Sb,Bi)_2$-based n-type Zintls. *J. Am. Chem. Soc.*, 140(5), 1910–1915.

Shuai, J., Tan, X.J., Guo, Q., Xu, J.T., Gellé, A., Gautier, R., Halet, J.F., Failamani, F., Jiang, J., Mori, T. (2019). Enhanced thermoelectric performance through crystal field engineering in transition metal-doped GeTe. *Mater. Today Phys.*, 9, 100094.

Song, L., Blichfeld, A.B., Zhang, J., Kasai, H., Iversena, B.B. (2018). Enhanced thermoelectric performance and high-temperature thermal stability of p-type Ag-doped β-Zn_4Sb_3. *J. Mater. Chem. A*, 6, 4079–4087.

Song, S., Mao, J., Bordelon, M., He, R., Wang, Y., Shuai, J., Sun, J., Lei, X., Ren, Z., Chen, S. et al. (2019). Joint effect of magnesium and yttrium on enhancing thermoelectric properties of n-type Zintl $Mg_{3+\delta}Y_{0.02}Sb_{1.5}Bi_{0.5}$. *Mater. Today Phys.*, 8, 25–33.

Sun, W., Sui, R., Yuan, G., Zheng, H., Zeng, Z., Xie, P., Yuan, L., Ren, Z., Cai, F., Zhang, Q. (2021). Thermoelectric module design to improve lifetime and output power density. *Mater. Today Phys.*, 18, 100391.

Tamaki, H., Sato, H.K., Kanno, T. (2016). Isotropic conduction network and defect chemistry in $Mg_{3+\delta}Sb_2$-based layered Zintl compounds with high thermoelectric performance. *Adv. Mater.*, 28(46), 10182–10187.

Tang, D., Zhu, W., Wei, P., Zhou, H., Liu, Z., Yu, J., Zhao, W. (2015). Preparation and properties of $Zn_4Sb_{2.94}In_{0.06}$/ZnO composite thermoelectric materials. *J. Electron. Mater.*, 44, 1902–1908.

Tewolde, M., Fu, G., Hwang, D.J., Zuo, L., Sampath, S., Longtin, J.P. (2015). Thermoelectric device fabrication using thermal spray and laser micromachining. *J. Therm. Spray Technol.*, 25(3), 431–440.

Wang, S.Y., She, X.Y., Zheng, G., Fu, F., Li, H., Tang, X.F. (2012). Enhanced thermoelectric performance and thermal stability in b-Zn_4Sb_3 by slight Pb-doping. *J. Electron. Mater.*, 41, 1091–1099.

Wang, X., Veremchuk, I., Burkhardt, U., Bobnar, M., Bottner, H., Kuo, C.-Y., Chen, C.-T., Chang, C.-F., Zhao, J.-T., Grin, Y. (2018). Thermoelectric stability of Eu- and Na-substituted PbTe. *J. Mater. Chem. C.*, 6, 9482–9493

Weitzman, L.H. (1967). Etching bismuth telluride. Patent, US3338765A.

Wood, M., Kuo, J.J., Imasato, K., Snyder, G.J. (2019). Improvement of low-temperature *zT* in a Mg_3Sb_2 -Mg_3Bi_2 solid solution via Mg-vapor annealing. *Adv. Mater.*, 31(35), 1902337.

Wood, M., Imasato, K., Anand, S., Yang, J., Snyder, G.J. (2020). The importance of the Mg–Mg interaction in Mg_3Sb_2-Mg_3Bi_2 shown through cation site alloying. *J. Mater. Chem. A*, 8(4), 2033–2038.

Wu, Y., Ming, T., Li, X., Pan, T., Peng, K., Luo, X. (2014). Numerical simulations on the temperature gradient and thermal stress of a thermoelectric power generator. *Energy Convers. Manag.*, 88, 915–927.

Xia, X., Huang, X., Li, X., Gu, M., Qiu, P., Liao, J., Tang, Y., Bai, S., Chen, L. (2014). Preparation and structural evolution of Mo/SiO_x protective coating on $CoSb_3$-based filled skutterudite thermoelectric material. *J. Alloys Compd.*, 604, 94–99.

Xing, T., Song, Q., Qiu, P., Zhang, Q., Gu, M., Xia, X., Liao, J., Shi, X., Chen, L. (2021). High efficiency GeTe-based materials and modules for thermoelectric power generation. *Energy Environ. Sci.*, 14(2), 995–1003.

Xu, C., Liang, Z., Shang, H., Wang, D., Wang, H., Ding, F., Mao, J., Ren, Z. (2021). Scalable synthesis of n-type $Mg_3Sb_{2-x}Bi_x$ for thermoelectric applications. *Mater. Today Phys.*, 17, 100336.

Xue, L., Zhang, Z., Shen, W., Ma, H., Zhang, Y., Fang, C., Jia, X. (2019). Thermoelectric performance of Cu_2Se bulk materials by hightemperature and high-pressure synthesis. *J. Materiomics*, 5, 103–110.

Yang, C.L., Lai, H.J., Hwang, J.D., Chuang, T.H. (2013). Diffusion soldering of $Bi_{0.5}Sb_{1.5}Te_3$ thermoelectric material with Cu electrode. *J. Mater. Eng.*, 22(7), 2029–2037.

Yang, J., Liu, G., Shi, Z., Lin, J., Ma, X., Xu, Z., Qiao, G. (2017a). An insight into β-Zn_4Sb_3 from its crystal structure, thermoelectric performance, thermal stability and graded material. *Mater. Today Energy*, 3, 72–83.

Yang, J., Zhang, X., Ge, B., Yan, J., Liu, G., Shi, Z., Qiao, G. (2017b). Effect of Zn migration on the thermoelectric properties of Zn_4Sb_3 material. *Ceram. Int.*, 43, 15275–15280.

Yin, H., Christensen, M., Pedersen, B.L., Nishibori, E., Aoyagi, S., Iversen, B.B. (2010). Thermal stability of thermoelectric Zn_4Sb_3. *J. Electron. Mater.*, 39, 1957–1959.

Yin, L., Chen, C., Zhang, F., Li, X., Bai, F., Zhang, Z., Wang, X., Mao, J., Cao, F., Chen, X. et al. (2020). Reliable N-type $Mg_{3.2}Sb_{1.5}Bi_{0.49}Te_{0.01}$/304 stainless steel junction for thermoelectric applications. *Acta Mater.*, 198, 25–34.

Ying, P., He, R., Mao, J., Zhang, Q., Reith, H., Sui, J., Ren, Z., Nielsch, K., Schierning, G. (2021). Towards tellurium-free thermoelectric modules for power generation from low-grade heat. *Nat. Commun.*, 12(1), 1121.

Zhang, J.F., Yang, K., Deng, C.M., Deng, C.G., Liu, M., Dai, M.J., Zhou, K.S. (2014). Method for preparing thick nickel coating on surface of semiconductor material. Patent, CN104357784A.

Zhang, Q.H., Huang, X.Y., Bai, S.Q., Shi, X., Uher, C., Chen, L.D. (2016). Thermoelectric devices for power generation: Recent progress and future challenges. *Adv. Eng. Mater.*, 18(2), 194–213.

Zhang, J., Song, L., Pedersen, S.H., Yin, H., Hung, L.T., Iversen, B.B. (2017). Discovery of high-performance low-cost n-type Mg_3Sb_2-based thermoelectric materials with multi-valley conduction bands. *Nat. Commun.*, 8, 13901.

Zhang, J., Song, L., Iversen, B.B. (2019). Insights into the design of thermoelectric Mg_3Sb_2 and its analogs by combining theory and experiment. *Npj Comput. Mater.*, 5(1), 76.

Zhang, X., Bu, Z., Lin, S., Chen, Z., Li, W., Pei, Y. (2020a). GeTe thermoelectrics. *Joule*, 4(5), 986–1003.

Zhang, Z., Zhao, K., Wei, T.-R., Qiu, P., Chen, L., Shi, X. (2020b). Cu_2Se-based liquid-like thermoelectric materials: Looking back and stepping forward. *Energy Environ. Sci.*, 13, 3307–3329.

Zhao, D. and Tan, G. (2014). A review of thermoelectric cooling: Materials, modeling and applications. *Appl. Therm. Eng.*, 66(1–2), 15–24.

Zhao, D., Geng, H., Chen, L. (2012). Microstructure contact studies for skutterudite thermoelectric devices. *Int. J. Appl. Ceram. Technol.*, 9(4), 733–741.

Zheng, Z., Su, X., Deng, R., Stoumpos, C., Xie, H., Liu, W., Yan, Y., Hao, S., Uher, C., Wolverton, C. et al. (2018). Rhombohedral to cubic conversion of GeTe via MnTe alloying leads to ultralow thermal conductivity, electronic band convergence, and high thermoelectric performance. *J. Am. Chem. Soc.*, 140(7), 2673–2686.

Zhu, Q., Song, S., Zhu, H., Ren, Z. (2019). Realizing high conversion efficiency of Mg3Sb2-based thermoelectric materials. *J. Power Sources*, 414, 393–400.

Zou, T., Xie, W., Feng, J., Qin, X., Weidenkaff, A. (2015). Recent developments in β-Zn_4Sb_3 based thermoelectric compounds. *J. Nanomater.*, 642909.

2

Effect of Microstructure in Understanding the Electronic Properties of Complex Materials

Chenguang FU, Chaoliang HU, Qi ZHANG, Airan LI and Tiejun ZHU
State Key Laboratory of Silicon Materials, School of Materials Science and Engineering, Zhejiang University, Hangzhou, China

2.1. Introduction

The conversion efficiency of a thermoelectric (TE) device depends on the Carnot efficiency and the TE materials' dimensionless figure of merit zT, $zT = S^2\sigma T/(\kappa_e+\kappa_L)$, where S, σ, T, κ_e and κ_L are the Seebeck coefficient, electrical conductivity, absolute temperature, electronic and lattice thermal components of total thermal conductivity κ, respectively. The S, σ and κ_e are intercoupled and mainly related to the charge carrier transport behavior, while κ_L is relatively independent and dominated by the phonon transport behavior. Hence, many microstructure engineering strategies, such as the introduction of atomic-scale point defects, nanoscale second-phases, mesoscale dislocations, textures and grain boundaries, have been widely used to suppress phonon transport, aiming at the minimization of κ_L and the improvement of TE performance. The mean free path (MFP) of phonons has

For a color version of all the figures in this chapter, see www.iste.co.uk/akinaga/thermoelectric2.zip.

Thermoelectric Micro/Nano Generators 2,
coordinated by Hiroyuki AKINAGA, Atsuko KOSUGA,
Takao MORI and Gustavo ARDILA.

a strong dependence on the frequency, and the induced microstructure in different length scales could hinder the propagation of phonons with various MFPs (Chen 2006). This is why microstructure engineering strategies are generally effective and popularly used in suppressing κ_L (Biswas 2012; Hu 2012). In contrast, the effect of microstructure engineering on the electronic properties of TE materials is more complicated and less studied. The electron can carry both heat and electric charge. Therefore, the electronic parameters S, σ and κ_e could also be affected when using the microstructure engineering strategies. A very common case is that the carrier mobility μ is deteriorated due to the introduced multi-scale scattering sources, which may reduce the gain to zT from the suppressed κ_L. In some other cases, the introduced scattering sources may be beneficial for S if they could scatter the low-energy charge carriers.

In section 2.2 of this chapter, we will first summarize the basic principles about electronic transport parameters and discuss how to gain insight into the carrier transport mechanisms from the experimental data of S, σ and μ. Then, in section 2.3, we will present several examples concerning the effect of point defects on electronic properties by focusing on isoelectronic alloying and intrinsic defects. Meanwhile, the effect of grain boundary and texture on electronic properties will also be elaborated. To conclude, in section 2.4, we will briefly highlight the future direction to gain a deeper understanding of the relationship between microstructure and electronic properties to minimize its negative effect on the TE properties.

2.2. Basic principles of electronic transport parameters

The electronic transport properties of heavily doped TE semiconductors could be understood using the Boltzmann transport equation under the relaxation time approximation. A detailed derivation of the expressions can be found in some classic books on thermoelectricity (Slack 1995; Goldsmid 2010). Here, the expressions obtained under the single parabolic band (SPB) model are given (Snyder 2012), which have been justified to be useful and feasible for understanding the electronic transport behavior of many bulk TE materials

$$n = \frac{4\pi(2m_d^* k_B T)^{\frac{3}{2}}}{h^3} F_{1/2}(\eta) \qquad [2.1]$$

$$\sigma = \frac{16\pi e^2}{3h^3} N_V (2m_b^*)^{\frac{1}{2}} (r + 3/2)(k_B T)^{\frac{3}{2}} \tau_0 F_{r+1/2}(\eta) \quad [2.2]$$

$$\mu = \frac{\sigma}{ne} = \frac{2e(r+3/2)\tau_0}{3m_b^*} \frac{F_{r+1/2}(\eta)}{F_{1/2}(\eta)} \quad [2.3]$$

$$S = \pm \frac{k_B}{e} \left[\eta - \frac{(r+5/2)F_{r+3/2}(\eta)}{(r+3/2)F_{r+1/2}(\eta)} \right] \quad [2.4]$$

$$L = \frac{k_B^2}{e^2} \left[\frac{r+7/2}{r+3/2} \cdot \frac{F_{r+5/2}(\eta)}{F_{r+1/2}(\eta)} - \left(\frac{r+5/2}{r+3/2} \right)^2 \cdot \frac{F_{r+3/2}^2(\eta)}{F_{r+1/2}^2(\eta)} \right] \quad [2.5]$$

$$F_i(\eta) = \int_0^\infty \frac{\xi^i}{\exp(\xi - \eta) + 1} \mathrm{d}\xi \quad [2.6]$$

$$\kappa_e = L\sigma T \quad [2.7]$$

where n is the carrier concentration, L is the Lorenz number, m_d^* is the density of states effective mass, N_V is the valley degeneracy, m_b^* is the single band effective mass ($m_d^* = N_V^{2/3} m_b^*$), e is the unit charge, h is the Planck constant, k_B is the Boltzmann constant, τ_0 is the relaxation time, r is the scattering factor, η is the reduced Fermi level (E_F/k_BT) and ξ is the reduced energy level (E/k_BT).

Based on equations [2.1]–[2.7], we can easily find that the electronic transport properties are closely related to m_b^*, η, r and τ_0. These four parameters could be divided into two groups: the band-structure-related m_b^* and η, and the scattering-related r and τ_0. Concerning the effect of microstructure on electronic transport properties, we could understand it through the change of either the band-structure or scattering mechanism:

1) m_b^* is inversely related to the curvature of an electronic band. Chemical doping and the formation of solid solutions may change m_b^* and thus affect the electronic properties. Other microstructure strategies, such as grain boundaries, nano-composites, dislocations, may not significantly change m_b^*.

2) η reflects the position of the Fermi level with regard to the edge of conduction or valence bands. Generally, the m_b^* of a single parabolic band does not change with the tuning of the Fermi level, and the value of η directly determines the concentration of charge carriers according to equation [2.1]. Chemical doping and the tuning of native defects could change η.

For carrier transport, the dominant scattering mechanism may vary at different temperatures and carrier concentrations. Before discussing how to gain insight into the scattering mechanism, the expressions of τ_0 or μ for the commonly occuring scattering processes in TE materials are first given below:

Acoustic phonon scattering (Bardeen and Shockley 1950):

$$\tau_{ac} = \frac{\pi\hbar^4 C_l}{2^{1/2} m_b^{*\,3/2} (k_B T)^{3/2} E_{def}^2} \xi^{-1/2} \quad [2.8]$$

where $\hbar$ is the reduced Planck constant ($h/2\pi$), C_l is the longitudinal elastic constant and E_{def} is the deformation potential, measuring the electron–phonon interaction potential.

Alloy scattering (Harrison and Hauser 1976a):

$$\tau_{al} = \frac{8\hbar^4}{3\pi\Omega x(1-x) m_b^{*\,3/2} (2k_B T)^{1/2} E_{al}^2} \xi^{-1/2} \quad [2.9]$$

where Ω is the volume per atom, x is the concentration ratio of the alloy atom and E_{al} is the alloy scattering potential which determines the magnitude of the alloy scattering for the given alloy.

Polar optical phonon scattering (Ravich and Morgovskii 1970):

$$\tau_{po} = \frac{4\pi\hbar^2}{e^2 m_b^{*\,1/2} (2k_B T)^{1/2} (\varepsilon_\infty^{-1} - \varepsilon_0^{-1})} \times [1 - \delta \ln(1 + \frac{1}{\delta})]^{-1} \xi^{1/2} \quad [2.10]$$

where ε_∞ and ε_0 are the high frequency and static dielectric constants, respectively. δ is a function of reduced carrier energy ξ defined as:

$$\delta(\xi) = \frac{e^2 m_b^{*\,1/2} N_V^{2/3}}{2(2k_B T)^{1/2} \xi \pi \hbar \varepsilon_\infty} F_{-1/2}(\xi) \quad [2.11]$$

Ionized impurity scattering (Harrison and Hauser 1976b):

$$\tau_{ii} = \frac{16\pi\varepsilon_0^2 (2m_b^*)^{1/2} (k_B T)^{3/2}}{N_{ii} Z^2 e^4} \times [\ln(1 + b) - \frac{b}{1+b}]^{-1} \xi^{3/2} \quad [2.12]$$

where N_{ii} is the concentration of ionized impurity and Z is the unit charge carried by each impurity. b is a function of the screening radius r_s: $b = 8m_b^*Er_s^2/\hbar^2$.

Grain boundary scattering (Seto 1975):

$$\mu = de\left(\frac{1}{2\pi m_b^* k_B T}\right)^{1/2} exp\left(-\frac{E_b}{k_B T}\right) \qquad [2.13]$$

where d is the grain size and E_b is the potential barrier height at grain boundaries.

Scattering mechanism	$\tau \propto E^r$	$\mu(T) \propto T^s$	
		Non-degenerate	*Degenerate*
Acoustic phonon	$E^{-1/2}$	$T^{-3/2}$	T^{1}
Alloy scattering	$E^{-1/2}$	$T^{-1/2}$	T^{0}
Ionized impurity	$E^{3/2}$	$T^{3/2}$	T^{0}
Polar optical phonon	$E^{1/2}$	$T^{-3/2}$	T^{1}

Table 2.1. *The energy dependence of relaxation time, as well as the temperature dependence of carrier mobility under the non-degenerate and degenerate limits, respectively, for several scattering mechanisms that commonly occurred in TE semiconductors*

The energy-dependent relaxation time and temperature-dependent carrier mobility for several scattering mechanisms that commonly occurred in TE semiconductors are summarized in Table 2.1. We can find that the exponents r and s vary for different scattering mechanisms, providing a way to judge the dominant scattering mechanism of carrier transport. Because the energy-dependent relaxation time is difficult to be characterized through transport measurements, the dominant carrier scattering mechanism could be judged using the $\mu \sim T^s$ relationship. As shown in Table 2.1, μ decays rapidly with increasing T with an exponent s of -1.5 (non-degenerate case) or -1 (degenerate case) if the acoustic phonon scattering dominates the carrier transport. This is commonly observed in the carrier transport of good TE semiconductors above room temperature. If the acoustic phonon scattering dominates the carrier transport, the scattering factor r is -1/2, according to equation [2.8]. With the confirmed value of r, we can thus understand the electronic properties through the SPB modeling analysis using equations [2.1]–[2.7]. For TE materials with isoelectronic alloying or high-content doping, the alloy scattering may contribute significantly to the carrier transport. In such a condition, the $\mu(T)$ will decay slower with rising T, and the exponent s is -0.5 or 0 for the non-degenerate and degenerate cases, respectively. Luckily, the scattering factor r stays the same as that of the acoustic phonon scattering. Hence, we can still use the SPB model to study electronic properties.

However, under the conditions of other scattering mechanisms, such as grain boundary and ionized impurity, r may change significantly, leading to much different σ, μ, S, κ_e. For lightly doped semiconductors, the ionized impurities may dominate the carrier transport at low temperatures, giving an r of 3/2. This indicates that high-energy carriers will be weakly hindered by the ionized impurities and could have a larger relaxation time compared to the low-energy ones. Under the ionized impurity scattering, μ will increase rapidly with rising T and s is 3/2 and 0 for the non-degenerate and degenerate cases, respectively. Polar optical scattering may contribute to the carrier transport for TE materials with the ionic bonding interaction. When it is the dominant one, r is 1/2, indicating an increased μ if n increases. This is the point to distinguish it from acoustic phonon scattering, as they show the same s value.

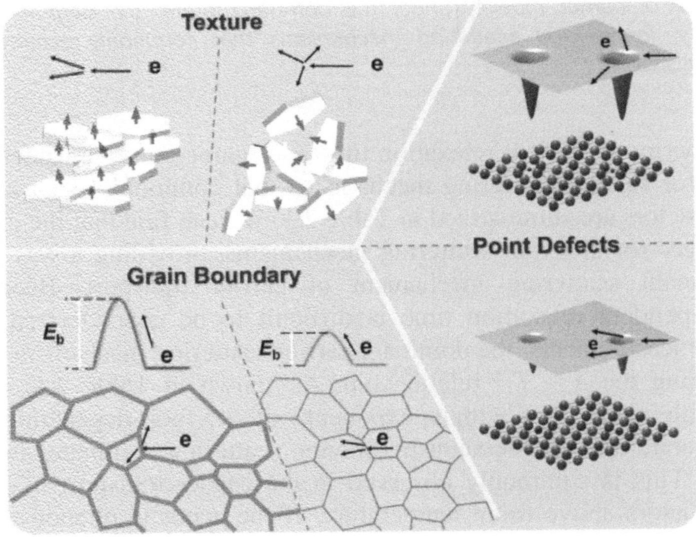

Figure 2.1. *A schematic summary showing the effect of point defects, grain boundary and texture on the carrier transport and how to minimize their negative effects. The right panel (point defects) was taken from the reference (Liu et al. 2018), copyright @ 2018, John Wiley and Sons*

Distinct from the above four scattering mechanisms, there is no fixed r or s for the grain boundary scattering. In some cases, mixed scattering mechanisms can also occur in the TE materials, which makes the analysis of their carrier transport behavior more difficult. For instance, there has been a recent debate on the carrier transport mechanisms in some polycrystalline TE materials near room temperature, i.e. whether it is dominated by impurity scattering or grain boundary scattering (Imasato et al. 2020; Li et al. 2020). In the next section, we will exemplify some experimental cases and discuss the effect of point defects, grain boundary and texture on electronic properties, and offer feasible strategies to realize enhanced TE performance, as summarized in Figure 2.1.

2.2.1. *Solid solutions*

The formation of solid solutions (i.e. alloying), initially proposed by Ioffe (1957), has long been employed in suppressing the κ_L of TE materials through enhancing the scattering of high-frequency phonons. However, the introduced alloying disorder may also scatter the charge carriers. To benefit the TE performance, a large value of μ/κ_L is desired when using this strategy. Aiming at a lower κ_L, an alloying element that has large mass and strain contrasts with the host elements is necessary. Meanwhile, to maintain μ, the alloying element should not introduce a strong perturbation to the periodic potential. Empirically, Ioffe made an argument on the selection of the alloying elements, i.e. an alloying element that makes less contribution to the charge-conducting band should be chosen (Wang et al. 2015). Through quantitative transport models, an instructive criterion that the alloying element should have small alloy scattering potential E_{al}, as well as large mass and strain contrasts, was proposed by Wang et al. (2013). The question then becomes how to choose an alloying element that could generate a small E_{al}. Several theoretical models were proposed to estimate E_{al}, i.e. considering the differences in the energy bandgap (Makowski and Glicksman 1973), the electron affinity (Harrison and Hauser 1976a) and the electronegativity (Ferry 1978). Based on the electronegativity theory, the difference in the covalent radius between alloying and host elements can be viewed as an indicator of E_{al}, offering an intuitive way to choose the alloying element.

Experimentally, Liu et al. (2018) demonstrated that lanthanide contraction serves as an indicator for selecting alloying elements. Owing to the lanthanide contraction, the radii of the period-6 transition metals are smaller than expected, which exhibit a similar value to the corresponding isoelectronic period-5 elements. Thus, the formation of solid solutions using isoelectronic period-5 and period-6 elements could generate a large mass contrast, but a small strain contrast, which could effectively suppress the κ_L while maintaining the μ (schematically shown in Figure 2.1). As experimentally demonstrated in the half-Heusler TE systems (Figure 2.2), the isoelectronic alloying of Zr/Hf or Nb/Ta weakly changes the lattice parameters, leading to a slight change in μ. In comparison, there is a large change in the lattice parameter of the n-type $PbTe_{1-x}Se_x$ system, which leads to a strong alloy scattering of the carriers and thus largely reduces the μ. These examples highlight the importance of the selection of alloying elements in optimizing the TE properties of solid materials.

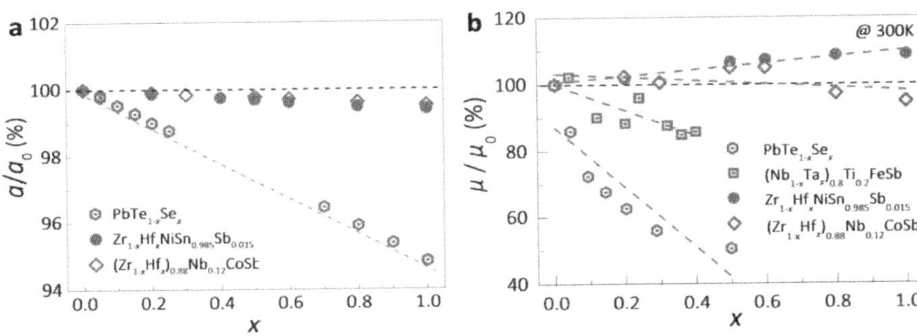

Figure 2.2. *(a) Normalized lattice parameter a and (b) normalized carrier mobility for several solid solution systems (Wang et al. 2013; Liu et al. 2015, 2018; Yu et al. 2018); a_0 and μ_0 denote the lattice parameter and carrier mobility of the unalloyed compound (x = 0), respectively*

2.2.2. *Intrinsic defects*

In the above-mentioned cases, the formation of solid solutions induces alloying disorder and thus affects both the μ and κ_L. But in some other TE materials, such as the Bi_2Te_3-based system, the formation of solid solutions (alloying Sb at the Bi site or Se at the Te site) will produce more complicated effects on the electronic and phonon transport properties, owing to the alloying-induced change of intrinsic point defects (Zhu et al. 2016). Although it is isoelectronic alloying in $Bi_{2-x}Sb_xTe_3$ and $Bi_2Te_{3-x}Se_x$, n and S can vary significantly with the increased content of the congeners, even changing the sign (Figure 2.3). It should be noted that the electronic properties of Bi_2Te_3-based single crystals will be discussed first. Bi_2Te_3 ingot is a p-type semiconductor resulting from the antisite defect $Bi_{Te}^{'}$ (Miller and Li 1965). The hole concentration increases by alloying Sb (Birkholz 1958; Starý et al. 1988; Ivanova and Granatkina 1995). This is because the formation energy of the antisite defect $Sb_{Te}^{'}$ is lower than that of $Bi_{Te}^{'}$, owing to the smaller differences in the electronegativity and covalent radius between the cation and anion (Starý et al. 1988; Zhu et al. 2016). On the contrary, if the differences increase, the formation energy of the antisite defects and anion vacancies will increase and decrease, respectively. The anion vacancies $V_{Se}^{\bullet\bullet}$ can provide two electrons and thus the Bi_2Se_3 ingot shows an n-type semiconducting behavior (West et al. 2012). Alloying Se at the Te site increases the formation energy of antisite defects and decreases the formation energy of anion vacancies. Therefore, the electron concentration increases and the *p–n* transition happens, as shown in the $Bi_2Te_{3-x}Se_x$ region (Figure 2.3) (Birkholz 1958; Horák et al. 1985; Lošt'ák et al. 2010).

However, the situation is different for polycrystalline $Bi_{2-x}Sb_xTe_3$-based alloys prepared by powder metallurgy (Figure 2.3) (Oh et al. 2000). The electron concentration increases for the fabricated bulk polycrystalline $Bi_{2-x}Sb_xTe_3$-based alloys compared to the single crystal counterparts, which was called the "donor-like effect" (Navrátil et al. 1996). This effect was explained by Navarátil et al. who propose a possible reaction mechanism between the vacancies and the antisite defects in the polycrystalline samples (for simplicity, we take Sb_2Te_3 as an example):

$$2V_{Sb}^{'''}+3V_{Te}^{\bullet\bullet}+Sb_{Te}^{'}\rightarrow V_{Sb}^{'''}+4V_{Te}^{\bullet\bullet}+Sb_{Sb}^{x}+6e^{'} \qquad [2.14]$$

The antisite defects are present in the melted ingots at relatively high concentrations. Subsequently, the cation and anion vacancies with a ratio of 2:3 are introduced by the non-basal slip during mechanical deformation (Schultz et al. 1962). The following hot-pressing sintering process will accelerate the reaction described in equation [2.14], resulting in the increase of electron concentration in the bulk polycrystalline samples. Thus, both mechanical deformation and the sintering process play vital roles in the generation of the donor-like effect (Zhang et al. 2019). Strong mechanical deformation and high sintering temperature will create more vacancies and accelerate the reaction of equation [2.14], respectively, facilitating the donor-like effect and thus increasing the electron concentration. On the other hand, changing the compositions can tune the concentration of antisite defects, thus affecting the donor-like effect and electron concentration (Hu et al. 2014). In short, changing the composition and the preparation method can effectively tune the intrinsic point defects and carrier concentration in Bi_2Te_3-based alloys and thus the TE properties. This could also happen in some other TE systems, particularly if their optimal carrier concentration is in the same order of magnitude as the concentration of the intrinsic defects.

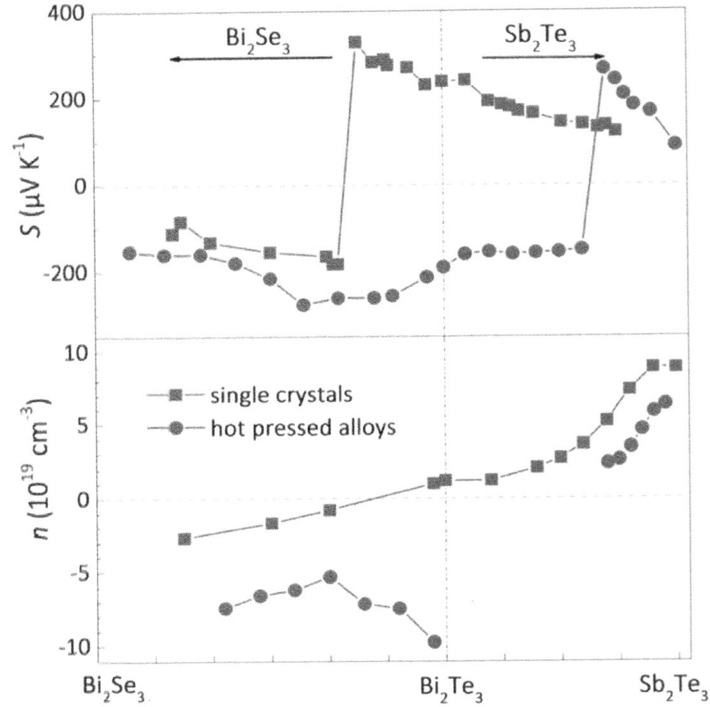

Figure 2.3. *The Seebeck coefficient and carrier concentration for $Bi_{2-x}Sb_xTe_3$ and $Bi_2Te_{3-x}Se_x$ systems, both in their single crystals and hot-pressed alloys (Birkholz 1958; Oh et al. 2000; Hu et al. 2014)*

2.2.3. *Grain boundary*

The grain refinement of bulk TE materials, usually introduced through the ball-milling method (Poudel et al. 2008), has been widely employed to enhance the grain boundary scattering (GBS) of phonons and suppress the κ_L of bulk TE materials in the past few decades. This bulk nanostructuring strategy was successfully used to improve the TE performance of a few TE materials (Minnich et al. 2009; Joshi et al. 2011; Yan et al. 2011; Dahal et al. 2014), such as, in the nanostructured $Si_{80}Ge_{20}P_2$, an obvious enhancement of the zT, which was achieved by Wang et al. (2008) (Figure 2.4(a)). Such an enhanced zT mainly results from the reduced κ_L, while the σ (Figure 2.4(b)) and μ are not significantly suppressed. These successes motivated extensive studies in the past decade to employ the nanostructuring strategy for the performance optimization of TE materials.

However, although grain refinement has shown its efficacy in optimizing the performance of a few TE materials, it has not always worked well for some other TE systems. As recently observed in the n-type Mg_3Sb_2 (Figure 2.4(c)), the grain refinement leads to a much lower zT in the polycrystalline sample near room temperature (RT), resulting from the largely reduced σ (Figure 2.4(d)) (Kanno et al. 2018). The rising σ with increasing T near RT incurs the debate on the carrier scattering mechanism, i.e. it is ionized impurity scattering or GBS. Although initially being recognized as the former, more evidence now supports that GBS is the dominant scattering mechanism near RT in the grain-refined polycrystalline samples, particularly when comparing the $\sigma(T)$ with that of single crystals. That is, the $\sigma(T)$ of n-type Te-doped Mg_3Sb_2 single crystals demonstrates the acoustic phonon scattering dominated carrier transport behavior (Imasato et al. 2020). It is worth noting that the n and S for n-type Mg_3Sb_2 samples are almost not affected by the grain refinement (Imasato et al. 2020). Moreover, owing to the intrinsically low κ_L and thus small phonon MFP, the phonon transport is also not significantly suppressed with increasing grain boundaries (Xin et al. 2018). Hence, for n-type Mg_3Sb_2, it is important to make large-size polycrystals or single crystals to achieve high room-temperature zT by eliminating the negative effect of GBS on the σ (Imasato et al. 2020). As observed from Figure 2.4(c), the average zT from 300 to 700 K displays an obvious increase in the large-size sample (Kanno et al. 2018).

By comparing the experimental results of n-type grain-refined $Si_{80}Ge_{20}$ and Mg_3Sb_2, we can find that whether the grain refinement is effective in enhancing the TE performance strongly depends on its effect on the carrier transport. According to equation [2.13], the potential barrier height E_b is a crucial parameter that determines whether GBS will significantly suppress the σ and μ (schematically shown in Figure 2.1). The E_b could be described as

$$E_b = \frac{e^2 Q_t^2}{8N\varepsilon_0} \qquad [2.15]$$

where e is the elementary charge (C), Q_t is the density of trapping states at grain boundaries (C / m^2), N is the concentration of ionized impurity atoms (usually dopant atoms, m^{-3}) and ε_0 is the static dielectric constant (C^2/N m^2).

It can be concluded that in the lightly doped samples with small ε_0, GBS is more likely to have a great effect on σ. The experimental evidence could be found in several cases: 1) for n-type Ir-doped NbFeSb, E_b decreases from 50 meV to 30 meV with increasing carrier concentration, resulting in higher μ (Shen et al. 2019); 2) for p-type PbSe, with the increased alloying content of $NaSbSe_2$, which has a lower ε_0 than that of PbSe, E_b was found to rise quickly from 1.6 meV to 68 meV, leading to deteriorative σ (Slade et al. 2020).

Since GBS may result in a serious deterioration of μ and σ, once GBS is confirmed to be the dominant carrier scattering mechanism, large-grain polycrystalline samples or single-crystalline samples are desired to exhibit better electronic properties and also higher TE performance. Generally, TE materials with intrinsically high κ_L, large ε_0 and higher n may benefit from the grain refinement strategy. In contrast, those with intrinsically low κ_L, small ε_0 and lower n may exhibit better TE performance if their grain sizes are larger.

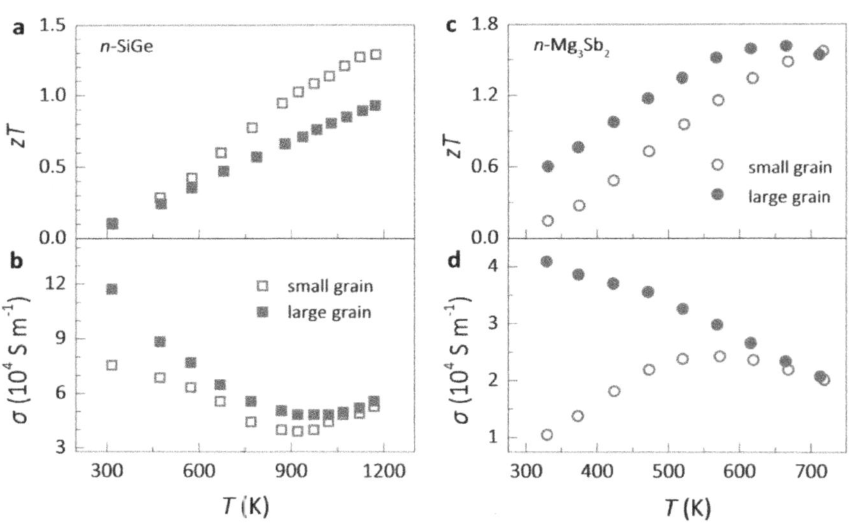

Figure 2.4. *The effect of grain boundaries on the zT and σ for n-type $Si_{80}Ge_{20}P_2$ and Mg_3Sb_2 polycrystalline samples (Wang et al. 2008; Kanno et al. 2018)*

2.2.4. *Texture*

Some TE materials, such as Bi_2Te_3, SnSe and BiCuSeO, show anisotropic electronic and phonon transport properties. That is, the TE performance in a specific direction is higher than in other directions. Taking n-type Bi_2Te_3 as an example, the σ in the in-plane (perpendicular to the c-axis) is about 4–7 times higher than that in the c-axis (Delves et al. 1961; Goldsmid 1961). In comparison, the κ_L in the in-plane direction is only twice higher than that of the out-of-plane (Shen et al. 2011). Consequently, the in-plane zT is higher than the out-of-plane zT. The Bi_2Te_3-based ingots prepared by the directional solidification are highly oriented polycrystals, which exhibit superior in-plane zT and are thus used for commercial TE modules. However, for polycrystalline Bi_2Te_3-based alloys prepared by powder metallurgy and hot-pressing (HP), the grains are randomly oriented in the bulk and the TE properties lose the anisotropy to a large extent, exhibiting deteriorated zT compared with the in-plane zT of single crystals (Pei et al. 2020).

Texturing is an effective method to improve the orientation of the polycrystalline samples, which could enhance the μ (schematically shown in Figure 2.1) and TE performance in the in-plane direction. Electron backscatter diffraction (EBSD) and X-ray diffraction (XRD) pole figures are useful methods to represent the texture of polycrystalline samples, which are thus often used to visually show the texture of polycrystalline Bi_2Te_3-based alloys (Hu et al. 2018; Wu et al. 2019). Besides, there is an easy and feasible way to characterize the texture, i.e. using the so-called orientation factor F of the (00l) plane. F is calculated by the Lotgering method according to the XRD results as follows (Lotgering 1959):

$$F = \frac{P-P_0}{1-P_0}, P_0 = \frac{\sum I_0(00l)}{\sum I_0(hkl)}, P = \frac{\sum I(00l)}{\sum I(hkl)} \quad [2.16]$$

where the P and P_0 are the integrated intensities of all (00l) planes to the (hkl) planes for the preferentially and randomly oriented samples, respectively. Some fabrication methods, such as the hot deformation (HD) (Hu et al. 2018) and liquid sintering (Yan et al. 2010; Kim et al. 2015) can facilitate the directional alignment of grains and enhance the texture of Bi_2Te_3-based alloys, as shown in Figure 2.5 (Wu et al. 2017). All the HD samples show higher F than the HP samples, indicating the increased texture of HD samples. The increased texture can also be visually observed from the fracture SEM images of the HP and HD samples (Wu et al. 2017). Due to the enhanced texture, the TE properties become more predominant in the direction perpendicular to the pressure. Therefore, the μ increases for the HD samples, as shown in Figure 2.5, which is favorable for a higher zT.

For TE materials that have non-cubic crystal structures, anisotropic transport properties in different crystallographic directions will occur. It is worth noting that whether the highest TE performance is achieved in the in-plane or the out-of-plane directions depends on the specific system. For instance, both p-type and n-type Bi_2Te_3-based alloys (Carle et al. 1995) show higher zT along the in-plane direction due to the higher μ, despite also having a higher κ_L. In comparison, the n-type SnSe (Chang et al. 2018) and $SnSe_2$ (Xu et al. 2017) show a higher zT along the out-of-plane direction, mainly contributed to the much lower κ_L. Thus, finding which direction exhibits higher TE properties is crucial for non-cubic TE systems. Based on this, the textured polycrystalline samples with the preferred direction can be fabricated to realize both good TE performance and mechanical strength.

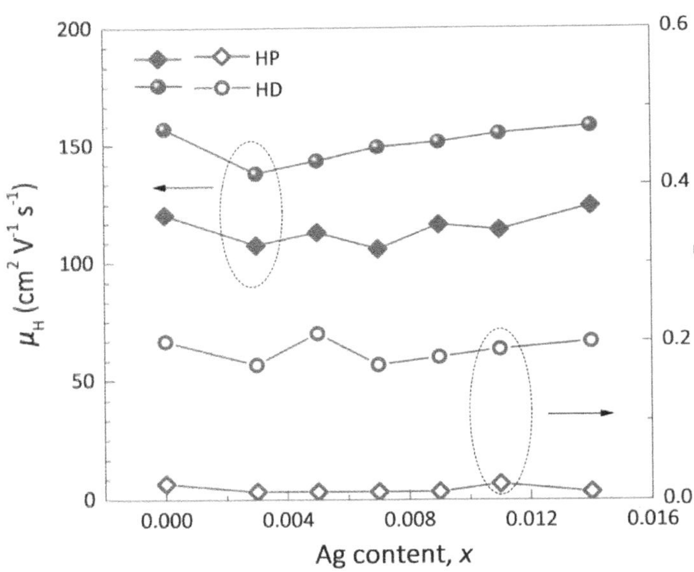

Figure 2.5. *Room-temperature Hall mobility and orientation factor F as a function of the Ag content in the HP and HD $Ag_xBi_2Te_{2.7}Se_{0.3}$ samples (Wu et al. 2017)*

2.3. Summary

Microstructure engineering, including the formation of solid solutions, native defects, grain refinement, texture and nanocomposites, have been widely used in TE studies, with the main purpose of the minimization of the lattice thermal conductivity. Since the MFP of phonons distributes in a broad range with respect to their frequencies, the microstructure engineering strategies are generally effective in suppressing the lattice thermal conductivity. However, the gain to the zT may be offset by the deteriorative carrier mobility and electrical conductivity. Thus, to benefit from microstructure engineering, its effect on electronic properties must be considered. Here, we exemplified several systems and discussed how to minimize the negative effect of point defects, grain boundary and texture on the electronic properties, aiming for the enhancement of TE performance. However, to the best of our knowledge, for some other types of microstructures, such as dislocations and nano-composites, their effects on electronic properties remain elusive and await future studies. Further, whether we can use microstructure engineering strategies to realize a simultaneous enhancement of electronic properties and suppress lattice thermal conductivity is an interesting issue waiting to be explored.

2.4. References

Bardeen, J. and Shockley, W. (1950). Deformation potentials and mobilities in non-polar crystals. *Physical Review*, 80(1), 72–80.

Birkholz, U. (1958). Untersuchung der intermetallischen Verbindung Bi2Te3 sowie der festen Lösungen $Bi_{2-x}Sb_xTe_3$ und $Bi_2Te_{3-x}Se_x$ hinsichtlich ihrer Eignung als Material für Halbleiter-Thermoelemente. *Zeitschrift für Naturforschung A*, 13(9), 780–792.

Biswas, K., He, J., Blum, I.D., Wu, C.I., Hogan, T.P., Seidman, D.N., Dravid, V.P., Kanatzidis, M.G. (2012). High-performance bulk thermoelectrics with all-scale hierarchical architectures. *Nature*, 489, 414.

Carle, M., Pierrat, P., Lahalle-Gravier, C., Scherrer, S., Scherrer, H. (1995). Transport properties of n-type $Bi_2(Te_{1-x}Se_x)_3$ single crystal solid solutions ($x \leqslant 0.05$). *Journal of Physics and Chemistry of Solids*, 56(2), 201–209.

Chang, C., Wu, M., He, D., Pei, Y., Wu, C.F., Wu, X., Yu, H., Zhu, F., Wang, K., Chen, Y. et al. (2018). 3D charge and 2D phonon transports leading to high out-of-plane ZT in n-type SnSe crystals. *Science*, 360(6390), 778–783.

Chen, G., and Dames, C. (2006). *Thermal Conductivity of Nanostructured Thermoelectric Materials*. CRC Press, Boca Raton.

Dahal, T., Jie, Q., Joshi, G., Chen, S., Guo, C., Lan, Y., Ren, Z. (2014). Thermoelectric property enhancement in Yb-doped n-type skutterudites $Yb_xCo_4Sb_{12}$. *Acta Materialia*, 75, 316–321.

Delves, R.T., Bowley, A.E., Hazelden, D.W., Goldsmid, H.J. (1961). Anisotropy of the electrical conductivity in bismuth telluride. *Proceedings of the Physical Society*, 78(5), 838.

Ferry, D.K. (1978). Alloy scattering in ternary III-V compounds. *Physical Review B*, 17(2), 912–913.

Goldsmid, H.J. (1961). Recent studies of bismuth telluride and its alloys. *Journal of Applied Physics*, 32(10), 2198–2202.

Goldsmid, H.J. (2010). *Introduction of Thermoelectricity*. Springer, Heidelberg.

Harrison, J.W. and Hauser, J.R. (1976a). Alloy scattering in ternary III-V compounds. *Physical Review B*, 13(12), 5347–5350.

Harrison, J.W. and Hauser, J.R. (1976b). Theoretical calculations of electron mobility in ternary III-V compounds. *Journal of Applied Physics*, 47(1), 292–300.

Horák, J., Lošťák, P., Koudelka, L., Novotný, R. (1985). Inversion of conductivity type in $Bi_2Te_{3-x}S_x$ crystals. *Solid State Communications*, 55(11), 1031–1034.

Hu, L., Gao, H., Liu, X., Xie, H., Shen, J., Zhu, T., Zhao. X. (2012). Enhancement in thermoelectric performance of bismuth telluride based alloys by multi-scale microstructural effects. *Journal of Materials Chemistry*, 22(32), 16484–16490.

Hu, L., Zhu, T., Liu, X., Zhao, X. (2014). Point defect engineering of high-performance bismuth-telluride-based thermoelectric materials. *Advanced Functional Materials*, 24(33), 5211–5218.

Hu, L., Zhang, Y., Wu, H., Liu, Y., Li, J., He, J., Ao, W., Liu, F., Pennycook, S.J., Zeng, X. (2018). Synergistic compositional–mechanical–thermal effects leading to a record high zT in n-type V_2VI_3 alloys through progressive hot deformation. *Advanced Functional Materials*, 28(35), 1803617.

Imasato, K., Fu, C., Pan, Y., Wood, M., Kuo, J.J., Felser, C., Snyder, G.J. (2020). Metallic *n*-type Mg_3Sb_2 single crystals demonstrate the absence of ionized impurity scattering and enhanced thermoelectric performance. *Advanced Materials*, 32(16), 1908218.

Ioffe, A.F. (1957). *Semiconductor Thermoelements and Thermoelectric Cooling*. Infosearch, London.

Ivanova, L.D. and Granatkina, Y.V. (1995). Properties of single-crystals in the Sb_2Te_3-Bi_2Te_3 solid-solution system. *Inorganic Materials*, 31(6), 678–681.

Joshi, G., Yan, X., Wang, H., Liu, W., Chen, G., Ren, Z. (2011). Enhancement in thermoelectric Figure-Of-Merit of an N-type half-Heusler compound by the nanocomposite approach. *Advanced Energy Materials*, 1(4), 643–647.

Kanno, T., Tamaki, H., Sato, H.K., Kang, S.D., Ohno, S., Imasato, K., Kuo, J.J., Snyder, G.J., Miyazaki, Y. (2018). Enhancement of average thermoelectric figure of merit by increasing the grain-size of $Mg_{3.2}Sb_{1.5}Bi_{0.49}Te_{0.01}$. *Applied Physics Letters*, 112(3), 033903.

Kim, S.I., Lee, K.H., Mun, H.A., Kim, H.S., Hwang, S.W., Roh, J.W., Yang, D.J., Shin, W.H., Li, X.S., Lee, Y.H. et al. (2015). Dense dislocation arrays embedded in grain boundaries for high-performance bulk thermoelectrics. *Science*, 348(6230), 109–114.

Li, A., Fu, C., Zhao, X., Zhu, T. (2020). High-performance $Mg_3Sb_{2-x}Bi_x$ thermoelectrics: Progress and perspective. *Research*, 1934848.

Liu, Y., Xie, H., Fu, C., Snyder, G.J., Zhao, X., Zhu, T. (2015). Demonstration of a phonon-glass electron-crystal strategy in (Hf,Zr)NiSn half-Heusler thermoelectric materials by alloying. *Journal of Materials Chemistry A*, 3(45), 22716–22722.

Liu, Y., Fu, C., Xia, K., Yu, J., Zhao, X., Pan, H., Felser, C., Zhu, T. (2018). Lanthanide contraction as a design factor for high-performance half-Heusler thermoelectric materials. *Advanced Materials*, 30(32), 1800881.

Lošt'ák, P., Drašar, Č., Bachan, D., Beneš, L., Krejčová, A. (2010). Defects in $Bi_2Te_{3-x}Se_x$ single crystals. *Radiation Effects and Defects in Solids*, 165(3), 211–215.

Lotgering, F.K. (1959). Topotactical reactions with ferrimagnetic oxides having hexagonal crystal structures I. *Journal of Inorganic and Nuclear Chemistry*, 9(2), 113–123.

Makowski, L. and Glicksman, M. (1973). Disorder scattering in solid solutions of III–V semiconducting compounds. *Journal of Physics and Chemistry of Solids*, 34(3), 487–492.

Miller, G.R. and Li, C.Y. (1965). Evidence for the existence of antistructure defects in bismuth telluride by density measurements. *Journal of Physics and Chemistry of Solids*, 26(1), 173–177.

Minnich, A.J., Lee, H., Wang, X.W., Joshi, G., Dresselhaus, M.S., Ren, Z.F., Chen, G., Vashaee, D. (2009). Modeling study of thermoelectric SiGe nanocomposites. *Physical Review B*, 80(15), 155327.

Navrátil, J., Starý, Z., PlecháčEk, T. (1996). Thermoelectric properties of p-type antimony bismuth telluride alloys prepared by cold pressing. *Materials Research Bulletin*, 31(12), 1559–1566.

Oh, T.S., Hyun, D.B., Kolomoets, N.V. (2000). Thermoelectric properties of the hot-pressed $(Bi,Sb)_2(Te,Se)_3$ alloys. *Scripta Materialia*, 42(9), 849–854.

Pei, J., Cai, B., Zhuang, H., Li, J. (2020). Bi2Te3-based applied thermoelectric materials: Research advances and new challenges. *National Science Review*, 7, 1856.

Poudel, B., Hao, Q., Ma, Y., Lan, Y., Minnich, A., Yu, B., Yan, X., Wang, D., Muto, A., Vashaee, D. et al. (2008). High-thermoelectric performance of nanostructured bismuth antimony telluride bulk alloys. *Science*, 320(5876), 634–638.

Ravich, Y.I. and Morgovskii, L.Y. (1970). Theory of scattering of carriers by optical and acoustic phonons in pbte-type semiconductors. *Soviet Physics Semiconductors*, 3, 1278.

Schultz, J.M., Mchugh, J.P., Tiller, W.A. (1962). Effects of heavy deformation and annealing on the electrical properties of Bi_2Te_3. *Journal of Applied Physics*, 33(8), 2443–2450.

Seto, J.Y.W. (1975). The electrical properties of polycrystalline silicon films. *Journal of Applied Physics*, 46(12), 5247–5254.

Shen, J., Hu, L., Zhu, T., Zhao, X. (2011). The texture related anisotropy of thermoelectric properties in bismuth telluride based polycrystalline alloys. *Applied Physics Letters*, 99(12), 124102.

Shen, J., Fan, L., Hu, C., Zhu, T., Xin, J., Fu, T., Zhao, D., Zhao, X. (2019). Enhanced thermoelectric performance in the n-type NbFeSb half-Heusler compound with heavy element Ir doping. *Materials Today Physics*, 8, 62–70.

Slack, G.A. (1995). *CRC Handbook of Thermoelectrics*. CRC Press, Boca Raton.

Slade, T.J., Grovogui, J.A., Kuo, J.J., Anand, S., Bailey, T.P., Wood, M., Uher, C., Snyder, G.J., Dravid, V.P., Kanatzidis, M.G. (2020). Understanding the thermally activated charge transport in $NaPb_mSbQ_{m+2}$ (Q = S, Se, Te) thermoelectrics: Weak dielectric screening leads to grain boundary dominated charge carrier scattering. *Energy & Environmental Science*, 13(5), 1509–1518.

Snyder, G.J. and May, A.F. (2012). *Introduction to Modeling Thermoelectric Transport at High Temperatures*. CRC Press, Boca Raton.

Starý, Z., Horák, J., Storceur, M., Stölzer, M. (1988). Antisite defects in $Sb_{2-x}Bi_xTe_3$ mixed crystals. *Journal of Physics and Chemistry of Solids*, 49(1), 29–34.

Wang, X.W., Lee, H., Lan, Y.C., Zhu, G.H., Joshi, G., Wang, D.Z., Yang, J., Muto, A.J., Tang, M.Y., Klatsky, J. et al. (2008). Enhanced thermoelectric figure of merit in nanostructured n-type silicon germanium bulk alloy. *Applied Physics Letters*, 93(19), 193121.

Wang, H., Lalonde, A.D., Pei, Y., Snyder, G.J. (2013). The criteria for beneficial disorder in thermoelectric solid solutions. *Advanced Functional Materials*, 23(12), 1586–1596.

Wang, H., Cao, X., Takagiwa, Y., Snyder, G.J. (2015). Higher mobility in bulk semiconductors by separating the dopants from the charge-conducting band – A case study of thermoelectric PbSe. *Materials Horizons*, 2(3), 323–329.

West, D., Sun, Y.Y., Wang, H., Bang, J., Zhang, S.B. (2012). Native defects in second-generation topological insulators: Effect of spin-orbit interaction on Bi_2Se_3. *Physical Review B*, 86, 121201(R).

Wu, Y., Zhai, R., Zhu, T., Zhao, X. (2017). Enhancing room temperature thermoelectric performance of n-type polycrystalline bismuth-telluride-based alloys via ag doping and hot deformation. *Materials Today Physics*, 2, 62–68.

Wu, Y., Yu, Y., Zhang, Q., Zhu, T., Zhai, R., Zhao, X. (2019). Liquid-phase hot deformation to enhance thermoelectric performance of n-type bismuth-telluride-based solid solutions. *Advanced Science*, 6(21), 1901702.

Xin, J., Li, G., Auffermann, G., Borrmann, H., Schnelle, W., Gooth, J., Zhao, X., Zhu, T., Felser, C., Fu, C. (2018). Growth and transport properties of Mg_3X_2 (X = Sb, Bi) single crystals. *Materials Today Physics*, 7, 61–68.

Xu, P., Fu, T., Xin, J., Liu, Y., Ying, P., Zhao, X., Pan, H., Zhu, T. (2017). Anisotropic thermoelectric properties of layered compound $SnSe_2$. *Science Bulletin*, 62(24), 1663–1668.

Yan, X., Poudel, B., Ma, Y., Liu, W.S., Joshi, G., Wang, H., Lan, Y., Wang, D., Chen, G., Ren, Z.F. (2010). Experimental studies on anisotropic thermoelectric properties and structures of n-type $Bi_2Te_{2.7}Se_{0.3}$. *Nano Letters*, 10(9), 3373–3378.

Yan, X., Joshi, G., Liu, W., Lan, Y., Wang, H., Lee, S., Simonson, J.W., Poon, S.J., Tritt, T.M., Chen, G. et al. (2011). Enhanced thermoelectric figure of merit of p-type half-Heuslers. *Nano Letters*, 11(2), 556–560.

Yu, J., Fu, C., Liu, Y., Xia, K., Aydemir, U., Chasapis, T.C., Snyder, G.J., Zhao, X., Zhu, T. (2018). Unique role of refractory Ta alloying in enhancing the figure of merit of NbFeSb thermoelectric materials. *Advanced Energy Materials*, 8(1), 1701313.

Zhang, Q., Gu, B., Wu, Y., Zhu, T.-J., Fang, T., Yang, Y., Liu, J., Ye, B., Zhao, X.-B. (2019). Evolution of the intrinsic point defects in bismuth telluride based thermoelectric materials. *ACS Applied Materials & Interfaces*, 11, 41424–41431.

Zhu, T., Hu, L., Zhao, X., He, J. (2016). New insights into intrinsic point defects in V_2VI_3 thermoelectric materials. *Advanced Science*, 3(7), 1600004.

3

Thermoelectric Nanowires

Olga CABALLERO-CALERO and Marisol MARTÍN-GONZÁLEZ
IMN-CNM, CSIC (CEI UAM+CSIC), Instituto de Micro y Nanotecnología, Isaac Newton, Tres Cantos, Madrid, Spain

3.1. Introduction

Among the different paths which can be followed to improve the properties of thermoelectric materials, the fabrication of nanoscale structures has been one of the most active. On the one hand, it is possible today to fabricate high-quality structures at the nanoscale of different thermoelectric materials. Techniques such as electrochemical deposition, vapor-liquid-solid processes, etc. are being used to obtain nanowires (NWs) with diameters ranging from tens to hundreds of nanometers and high aspect ratios. On the other hand, the nanowire structures can be approached as a one-dimensional structure for most transport phenomena, and thus, the obtained results can be successfully interpreted with one-dimensional theories, simplifying both explanations and predictions. Therefore, they are used as an important platform for understanding the influence of dimensionality reduction in thermoelectric materials. It is generally assumed now that the increased surface-to-volume ratio produced in nanowires increases the diffusive phonon scattering, which reduces the thermal conductivity. In these kinds of structures, the study of the influence of the surfaces on the phonon

For a color version of all the figures in this chapter, see www.iste.co.uk/akinaga/thermoelectric2.zip.

Thermoelectric Micro/Nano Generators 2,
coordinated by Hiroyuki AKINAGA, Atsuko KOSUGA,
Takao MORI and Gustavo ARDILA.

scattering, from the theoretical point of view, has helped us to further understand the thermal transport at the nanoscale. Recently, more sophisticated structures, based on nanowires and maintaining their enhanced properties for thermoelectric applications, have been implemented and tested.

In this chapter, we will present the different paths that are being studied to obtain relevant thermoelectric enhancements in nanowire structures and to what extent they have been developed. We will review the fabrication methods for obtaining thermoelectric nanowires, along with the measurement techniques and physical properties obtained in the literature. This will provide us with a way of understanding the physics behind the different ways that are being studied to enhance the thermoelectric efficiency in these structures. Then, we will briefly discuss the implementation of thermoelectric nanowires in actual devices with some examples. Finally, we will take a look at a new family of nanostructures based on the combination of nanowire structures in three-dimensional arrangements and their properties, concluding with some outlooks for future developments in this field.

3.2. Nanowires: a way to enhance thermoelectric efficiency

As previously discussed in this book, the thermoelectric performance of a certain material is related to its figure of merit, zT. In this parameter, different properties of the material (S, Seebeck coefficient, σ, electrical conductivity and κ thermal conductivity) are involved. Several strategies can be followed in NWs to improve the thermoelectric performance of materials (see Figure 3.1) which will be discussed here.

The figure of merit is proportional to the power factor, $S^2 \cdot \sigma$, and inversely proportional to its thermal conductivity. One of the main obstacles to increase the figure of merit has been that all parameters involved are interlinked in classical physics. Then, going to low-dimensional structures, such as nanowires, we can decouple those parameters and produce an enhancement of thermoelectric efficiency. In 1993, Dresselhaus predicted that in low-dimensional structures, the thermoelectric power factor would be drastically increased due to quantum confinement effects (Hicks and Dresselhaus 1993). To understand the actual effects that take place at the nanoscale, we should take into account the form that these parameters have using the Boltzmann transport equations. Then, σ and S are expressed as follows:

$$\sigma = \frac{e^2}{3} \int g(E) v^2(E) \tau(E,T) \left(-\frac{\partial f(E)}{\partial E}\right) dE \qquad [3.1]$$

$$S = \frac{k_B}{e} \left(\frac{\int_0^\infty g(E) v^2(E) \frac{E-E_F}{k_B T} \tau(E,T) \left(-\frac{\partial f(E)}{\partial E}\right) dE}{\int_0^\infty g(E) v^2(E) \tau(E,T) \left(-\frac{\partial f(E)}{\partial E}\right) dE} \right) \qquad [3.2]$$

In these equations, e is the electron charge, $g(E)$ is the total electronic density of states (DOS), $v(E)$ is the electron velocity at a given energy E, $\tau(E,T)$ is the carrier's relaxation time at a certain energy and temperature, $f(E)$ is the energy distribution function, and k_B is the Boltzmann constant. In the case of charge carriers, $f(E)$ is a Fermi–Dirac distribution, and its derivative with respect to energy has a peak at the Fermi level, E_F, ensuring that only the carriers close to that energy contribute to both the electrical and thermoelectrical transport. Also, $g(E)$ in semiconductors is proportional to the effective mass of the charge carrier of the DOS, m^*.

The increase of the carrier concentration (through doping, for instance) produces an increase in E_F and a higher amplitude of $g(E)$, increasing σ (see equation [3.1]). But it also causes a less symmetrical distribution of the function $g(E)$ around E_F, which reduces the value of S. Therefore, a way of increasing the power factor is to find the optimal carrier concentration to maximize both σ and S. The carrier's relaxation time, $\tau(E,T)$, is energy-dependent, and it is inversely proportional to the carrier mobility μ, and it also appears in equations [3.1] and [3.2], and thus, influences the power factor. If we assume $\tau(E) \sim E^r$, that is, a scattering law that goes as the energy E to a certain power, named r, and parabolic energy bands of the form $E = E_c \pm \hbar^2 k^2 / 2m^*$ (+ for electrons, − for holes), which can be assumed when reaching the quantum regime, the power factor can be written as follows:

$$S^2\sigma \propto \frac{E^r}{m^*} \left(\frac{E_F - E_C}{k_B T} + \left(r + \frac{5}{2}\right) \right)^2 \qquad [3.3]$$

Then, for higher r values, the power factor will be higher, which can be understood as a higher scattering of low energy carriers. Another effect that can enhance the power factor is to modify the DOS, which was the route proposed by Dresselhaus in 1993 (Hicks and Dresselhaus 1993). In this work, the authors proposed sharp DOS at low dimensions, along with quantum-confined sub-bands, and following this way of thought, the zT on these structures would be highly enhanced. These effects cannot be achieved in bulk materials, but the fact of reducing the dimensionality makes it possible to reach these phenomena. Nevertheless, the achievements of these effects in nanowires have not been proven experimentally, due to the extremely small diameters and high quality required, or the actual DOS shape, which is more complex than the theoretical approximations. But the research carried out in this sense has resulted in other ways of improving the *zT* in these structures, such as the Coulomb blockade effect, which is a charging effect that blocks the transport of electrons through the structure, found in InAs NWs (Wu et al. 2013) or the "carrier pocket engineering", also introduced by Dresselhaus (Koga et al. 1998), which consists of engineering the DOS in such a way that the valley degeneracy increases, forming quantum well states which can separate the carriers in different "pockets" in the real space, minimizing the inter-valley scattering, and thus, enhancing the power factor.

Another important strategy for increasing *zT* is the reduction of thermal conductivity. It is made up of the sum of two terms: one due to carriers and thus linked to transport properties like σ, and another due to phonons and lattice parameters, which can be engineered independently of the other transport properties. The first results in this sense were triggered by the research in NWs when trying to enhance the power factor, and it was found that, even though the dimensions needed for getting quantum effects were not easily achieved, the increase of the surface-to-volume ratio in these one-dimensional structures provided a way of reducing the thermal conductivity (by increasing the phonon scattering on the surface). For instance, enhanced figures of merit for silicon (*zT* of 0.01 at RT) when it was engineered in the form of nanowires of 20 nm in diameter were found up to 0.25 at RT (Boukai et al. 2008). Moreover, a *zT* of 0.6 at RT was achieved for silicon nanowires of 50 nm in diameter when the surfaces were rough, and thus, the phonon scattering was further increased (Hochbaum et al. 2008). Also, in the case of Bi_2Te_3 nanowires, a thorough study on the thermal conductivity decrease obtained when decreasing the diameter presented in 2017 showed that the phonon scattering, in this case, can be understood under the Kinetic Collective model (Rojo et al. 2017) in all the diameters studied in this work. In this sense, once the effect of the size reduction in phonon scattering processes is understood, further enhancement of phonon scattering in NWs has been achieved through devising complex morphologies (segmented or diameter modulated NWs, for instance) which will produce a reduction in the total thermal conductivity and, thus, an increase on *zT*.

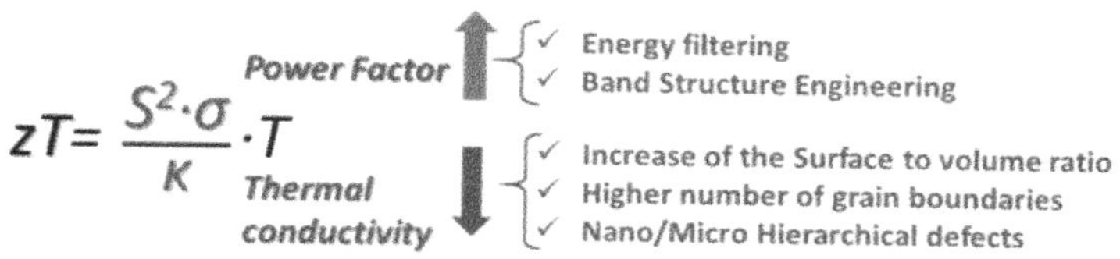

Figure 3.1. *Examples of the main strategies to increase the thermoelectric figure of Merit, zT, can be implemented in nanowires, being zT the thermoelectric figure of merit, S the Seebeck coefficient, σ the electrical conductivity, κ the thermal conductivity and T the absolute temperature*

3.3. Fabrication of thermoelectric nanowires

A wide variety of techniques have been used to fabricate thermoelectric nanowires, most of them being bottom-up approaches, but also some top-down methods, usually focused on the fabrication of silicon nanowires, such as electron beam lithography (Pennelli and Piotto 2006), nanosphere lithography (Hamdana et al. 2017) or atomic force lithography (Martinez et al. 2008) along with oxidation or etching processes (Mallavarapu et al. 2020), have been used in the literature. These methods, although widely used for the microfabrication of different structures, are usually more expensive than most bottom-up approaches for nanowire fabrication, although they provide better reproducibility as far as crystallinity, composition or doping are concerned, given that those properties can be tailored in the bulk and then, the nanowires engineered from that bulk will maintain those properties. Other top-down techniques that have been developed lately are metal-assisted-chemical-etching (MACE), which has proven more cost-efficient and suitable for large-scale production of vertical silicon nanowires (Elyamny et al. 2020) and porous silicon nanowires (Yang et al. 2021) or focused ion beam irradiation by the implantation of gallium and selective etching to obtain suspended silicon nanowires (Llobet et al. 2018).

Given that the most extended approaches are bottom-up (Zhu et al. 2021), we will discuss them more in-depth. Among the advantages of these techniques, we can count the possibility of large-scale fabrication at a lower cost, and the achievement of smaller structures than those obtained by top-down approaches. Nevertheless, it is not always possible to find a bottom-up technique producing the desired composition or crystalline orientation and posing questions in order to find the right chemical process to produce them have to be taken into account. Among these techniques vapor-liquid-solid (VLS) has been widely used to produce Si (Díez et al. 2020, Hochbaum et al. 2008), or SiGe nanowires (Fonseca et al. 2021), and sometimes, even chalcogenides such as Bi_2Te_3 (Park et al. 2016) or Bi_2Se_3 (Kong et al. 2010). VLS was proposed by Wagner and Ellis in 1964 while working at Bell Labs (Wagner and Ellis 1964), and consists of using metal catalytic nanoparticles on a silicon substrate, which are then exposed to a vapor phase precursor, such as silane or germane, of the nanowire material. Then, under each metallic nanoparticle the crystallization of the material is produced and a nanowire is grown, for instance, GaAs NWs (Ghukasyan et al. 2022). This method is highly compatible with silicon technology and the fabrication of horizontal arrays of silicon nanowires for their integration in thermoelectric devices has been performed with this technique (Dávila et al. 2012). Lately, the growth of certain nanowires, such as single-crystalline Ag_2Te, has been carried out via a vapor-transport method without the need for a catalyst (Pillai 2021). Chemical Vapor Deposition (CVD) has also been used to obtain a CdSSe nanowire-based chip (Ding et al. 2021). Another technique widely used for semiconductors, molecular beam epitaxy (MBE), using metal nanoparticles to promote the growth of the semiconductor nanowires, has been employed to grow thermoelectric nanowires of Ge (Lange et al. 2020) or Bi_4Te_3 (Wang et al. 2009) or Chemical Beam Epitaxy as for InAsSb nanowires (Peri et al. 2022).

Other families of bottom-up techniques are those where the growth is obtained from a solution (Zhu et al. 2021), such as the hydrothermal (the synthesis is done in water) or solvothermal (other solvents) methods, which has been used to obtain Sb_2Se_3 or Cu_2Se thermoelectric nanowires (Kim et al. 2021). Other solution-based techniques are the sonoelectrochemical processes (Qiu et al. 2005), template-assisted die-casting (Vivekanandan et al. 2022) or template-assisted electrochemical growth, which is the most widely used method for obtaining thermoelectric nanowires. Electrochemistry provides a scalable to industry, room temperature and cost-effective method for obtaining a wide variety of materials in a reproducible and controlled way. If ion-track polycarbonate membranes or anodic alumina templates, where a high density of pores in the range of tens to hundreds of nanometers in diameter and microns in length can be easily achieved, are used as templates, the electrochemical growth will fill those pores and produce nanowires with the corresponding geometry. However, other techniques, such as Atomic Layer Deposition (ALD) for ZnO nanowires (Ruoho et al. 2016), and Sb_2S_3 on VLS grown Sb_2Se_3 NWs (Yang et al. 2010) or vacuum melting and pressure injection methods (to produce Bi nanowires (Bisrat et al. 2007; Zhang et al. 1999) or Poly(3-hexylthiophene) – P3HT – nanowires, for instance (Rojo et al. 2014)) can also be used to fill such hollow membranes and produce nanowires. In the literature, we can mostly find template-assisted electrochemically produced thermoelectric nanowires: Bi_2Te_3 (Manzano et al. 2019) (Rojo et al. 2013b) and other chalcogenides from groups V–VI, as Sb_2Te_3 or ternary compounds BiTe:Sb or BiTe:Se (Manzano and Martín-González 2019) or single elements like Bi, Sb, etc. (Gorokh et al. 2017).

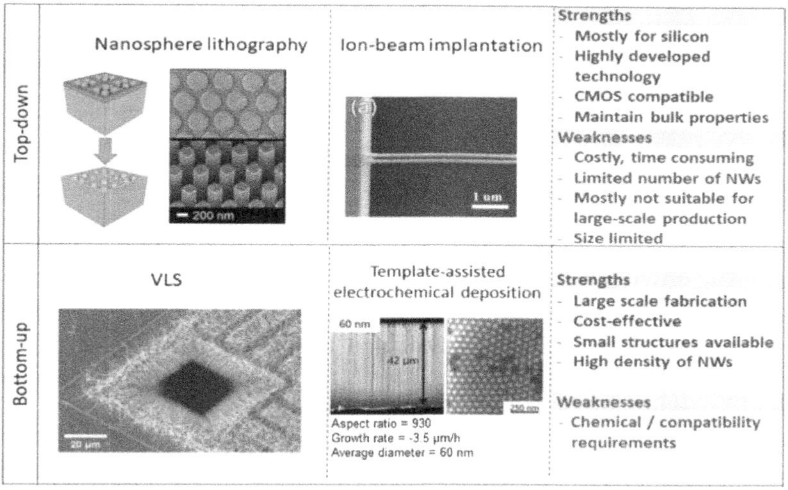

Figure 3.2. *Examples of different fabrication approaches of nanowires, along with their general advantages (in green) and drawbacks (red). The images have been reproduced with permission, starting from up left and in a clockwise direction (Hamdana et al. 2017), (Llobet et al. 2018), (Manzano et al. 2019) and (Fonseca et al. 2021)*

As a whole, there is a wide variety of techniques that, depending on the material or the properties needed, can be chosen to obtain the best nanowires for a given application. Figure 3.2 presents some of the possible nanowire configurations and properties associated with the most widely used techniques.

3.4. Measurement of thermoelectric properties in nanowires

There are different experimental techniques devised for the measurements of the thermoelectric properties of nanowires to characterize both the power factor and the thermal conductivity (see, for instance, Rojo et al. (2013a) for a detailed review on them). It is important to note that in most cases, when measuring nanowires we are aiming at the nanoscale, and even though nanotechnology is greatly developed nowadays, most of the measurements and understanding of the results remain an experimental challenge (Yang et al. 2022). Moreover, there is a lack of metrology in these measurements, and it is important to have different techniques to compare obtained results in different laboratories, as was done, for instance, in Rojo et al. (2017), where two different methods were used to characterize the thermal conductivity of Bi_2Te_3 nanowires. In most cases, the zT of the nanowires is obtained from separated measurements of the different properties, S, σ and κ.

In general, we can divide these techniques into two: (i) those who measure ensembles of nanowires as a whole, for instance, inside the templates in which they are grown, considering a composite material, or after pressing a bunch of nanowires into a pellet or thin film (Chen et al. 2022); and (ii) single nanowire measurements, developed with the aid of micro- and nanotechnologies, usually imply the fabrication of a micro-chip in which a single nanowire is individually connected and measured. Figure 3.3 shows schematically all these techniques along with their main advantages and drawbacks, which will be briefly discussed next.

Within the measurements carried out for an ensemble of nanowires, those usually employed when nanowires are grown inside a template, for instance. In that case, the nanowires embedded in the template are treated as a composite, and then the contribution of the matrix and the nanowires are separated. Examples of these measurements are thermal conductivity measurements with the photoacoustic technique, in which the effective medium theory is used to extract both contributions from the measurement (Abad et al. 2016, 2017; Biswas et al. 2009) or the measurement of GaAs NW arrays with the 3ω technique (Ghukasyan et al. 2022). In the case of electrical conductivity measurements, the main problem is that they are mostly two-probe measurements, and thus, have a great associated error. Finally, for the Seebeck coefficient measurement, it can be done by contacting electrically the top and the bottom part of the nanowires and maintaining a thermal gradient, obtaining the Seebeck from the thermal gradient established and the voltage obtained. An example of such measurements can be found in Cassinelli et al. (2017). Also in this category of measurements, we can include those performed in high numbers of individual nanowires pressed together in the form of pellets (Brockway et al. 2014) or thin films (Tai et al. 2008) and then treated as a composite material. In those cases, the techniques employed can be the same used for bulk or thin films, but several assumptions have to be made to obtain the contribution of the individual nanowires to the different parameters (for instance, the contact resistivity when measuring electrical conductivity).

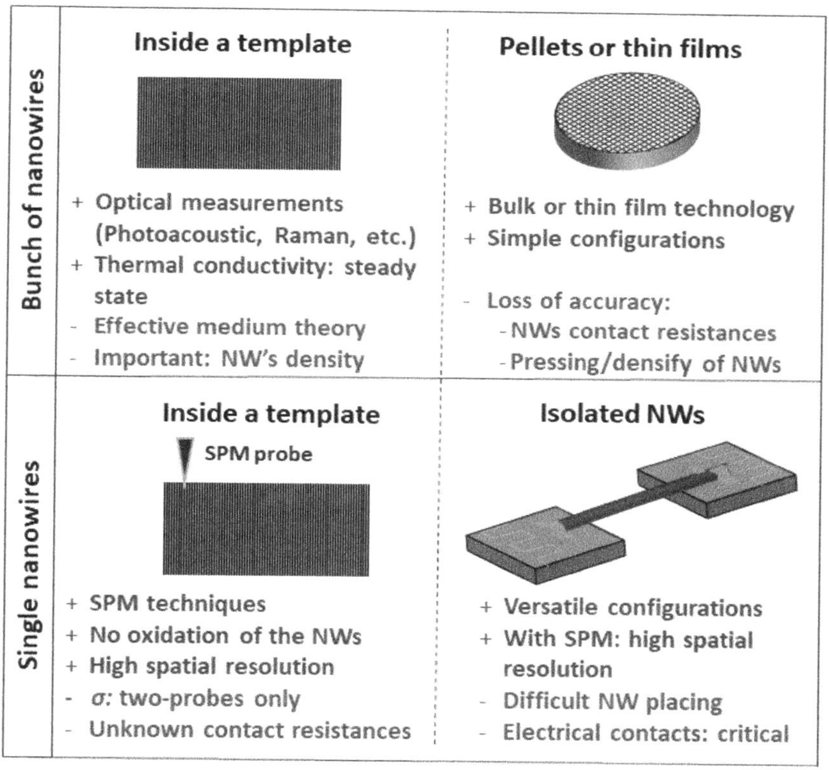

Figure 3.3. *Schematic summary of the different configurations used in NWs to measure the thermoelectric properties, either in ensembles or for single nanowires. Their general advantages are in green and drawbacks in red*

The characterization of single nanowires can be done either when those are still inside a template matrix, or for isolated nanowires, which in the case of template-assisted fabrication methods, can be obtained after dissolving the matrix. The first type of measurements in embeded nanowires can be performed using Scanning Probe Microscopies (SPM). In them, conductive SPM probes can be used to contact the end of a single nanowire at one side of the template, while the other end is electrically connected also, and make a standard two-probe σ measurement or thermoresistive probes which enable the characterization of the thermal conductivity of single nanowires, as it is done in 3ω-Scanning Thermal Microscopy (3ω-SThM) (Rojo et al. 2013b). In the case of SThM measurements of nanowires within a matrix, a model must be used to extract the value of the nanowires, since the tip exchange radius is normally higher than the diameter of the nanowires to be measured. The SThM technique to measure thermal conductivity at the nanoscale is becoming more popular since the SThM are now commercial and not lab-developed as they were previously. The second family of measurements on single nanowires are those based on micro-platforms designed for different properties (Yang et al. 2021). For instance, a single nanowire can be placed on a microchip and contacted at both ends (two or four-probe measurements) to characterize its electrical conductivity. If a temperature gradient is also created along the nanowire (with microheaters in the same chip), then the temperature drop can be measured with two electrodes which can also be used as thermometers, and thus, the Seebeck coefficient can be obtained (Mavrokefalos et al. 2009). Also, field-effect transistors can be used to modulate the carrier concentration in the nanowires by changing the gating, and thus, the power factor can be obtained (Sun et al. 2018; Fust et al. 2020). Suspended microdevices have also been employed to perform thermal conductivity measurements in nanowires, where two pads containing heaters and electrodes are bridged with the nanowire, which is then electrically contacted (Valentín et al. 2013) or by performing 3ω measurements in single nanowires (Peri et al. 2022). The main problem with these kinds of measurements is assessing an ohmic contact with the NWs, which is crucial to obtain reliable results. Another possibility is to use an SPM to scan the surface of the NW in the chip, and thus measure the electrical conductivity along its length, achieving high-spatial resolution (Rojo et al. 2016). Alternatively, Sojo-Gordillo et al. (2021) use the previously cited SThM technique to measure the thermal conductivity of suspended nanowires.

In the last few years, there have been results on different kinds of nanowires prepared by the different techniques discussed above. In the following table, we present a summary of the publications where the three parameters S, σ, κ have been measured in the nanowires and the zT is reported. Those values are visually summarized in Figure 3.4. Table 3.1 includes the material, NW diameter, growth technique, S, σ, κ, zT, values at room temperature along with the reference.

Material	Diameter (nm)	Growth technique	S (μV/K)	σ (S/m)	κ (W/m·K)	zT @ RT	*References*
Bi_2Te_3	8	Sol-pellets	-205	$5.1\cdot10^4$	1.42	0.45	(Zhang et al. 2012b)
Bi_2Te_3	< 20	Sol-pellets	-150	$4.6\cdot10^4$	0.45	0.69	(Zhou et al. 2016)
$Bi_2(Te,Se)_3$	20	Th-co-ev.	-189	$4.9\cdot10^4$	0.52	1.01	(Tan et al. 2014)
Bi_2Te_3	52	ED-AAO	-50	$21\cdot10^4$	2.9	0.06	(Mavrokefalos et al. 2009)
Bi_2Te_3	67	ED-AAO	-70	$16\cdot10^4$	2.3	0.1	(Shin et al. 2014)
Bi_2Te_3	67	ED-AAO (HT)	-100	$20\cdot10^4$	2.6	0.21	(Shin et al. 2014)
Bi_2Te_3	75	ED-AAO (Twins)	-57	$16\cdot10^4$	1.9	0.08	(Shin et al. 2014)
Bi_2Te_3	75	ED-AAO (Twins+HT)	-140	$7.5\cdot10^4$	1.4	0.31	(Shin et al. 2014)
Bi_2Te_3 (Te-rich)	120	ED-AAO	-67	$25\cdot10^4$	0.75	0.45	(Chen et al. 2010)
Bi_2Te_3	1000	VLS	40	$17\cdot10^4$	1.2	0.068	(Hsin et al. 2013)
3D-Bi_2Te_3	50	ED-3DAAO	-127	$7\cdot10^4$	0.501	0.55	(Ruiz-Clavijo et al. 2021)
Bi-Te core-shell	25	Sol-pellets	-135	$0.52\cdot10^4$	0.55	0.05	(Zhang et al. 2008)
Bi-Te core-shell	300	Bi NW+ Te sputtering	-100	$45\cdot10^4$	2	0.5	(Kim et al. 2018)
$(Sb,Bi)_2Te_3$	<20, tilt 90°	Thermal evap.	204	$5.1\cdot10^4$	0.71	0.9	(Tan et al. 2018)
$(Sb,Bi)_2Te_3$	<20 tilt 60°	Thermal evap.	221	$7.4\cdot10^4$	0.89	1.22	(Tan et al. 2018)
$(Sb,Bi)_2Te_3$	<20 tilt 45°	Thermal evap.	246	$8.8\cdot10^4$	0.93	1.72	(Tan et al. 2018)
$Bi_{0,5}Sb_{1,5}Te_3$	67	ED-AAO pulsed	125	$4.5\cdot10^4$	0.29	0.73	(Li et al. 2015)
SnTe:In	100-200	Chemical-vapor	60	$14.5\cdot10^4$	2.25	0.07	(Li et al. 2017)
SnSe	130	VLS	220	$20\cdot10^4$	0.55	0.005	(Hernandez et al. 2018)
InSb	80-200	Au-assist. chemBE	-306	$1.65\cdot10^4$	17.5	0.025	(Yazji et al. 2016)
InSb	170	VLS	-100	$2\cdot10^4$	5	0.005	(Zhou et al. 2009)
InAsSb	60-80	CBE SL	-180	$3\cdot10^4$	1.2	0.24	(Peri et al. 2022)

Material	Diameter (nm)	Growth technique	S (μV/K)	σ (S/m)	κ (W/m·K)	zT @ RT	*References*
Zn_4Sb_3	100	PVD	147	$4.5 \cdot 10^4$	0.55		
Te-BiTe barbell	37	Sol-pellets	610	$0.3 \cdot 10^4$	0.36	0.1	(Zhang et al. 2012a)
AgTe-BiTe barbell	25	Sol-pellets	287	$0.1 \cdot 10^4$	1.6	0.1	(Fang et al. 2014)
PbTe-BiTe barbell	30	Sol-pellets	-290	$4 \cdot 10^4$	0.57	0.18	(Fang et al. 2013)
PbTe-AgTe barbell	20	Sol-pellets	200	$0.18 \cdot 10^4$	0.28	0.077	(Yang et al. 2015)
PbTe	12	Sol-pellets	245	$1.5 \cdot 10^4$	1	0.27	(Finefrock et al. 2014)
PbTe	180	VLS	-72	$0.45 \cdot 10^4$	1.2	0.0054	(Lee et al. 2011)
PbTeSe	<20	Solution	-210	$2.15 \cdot 10^4$	0.52	0.6	(Zhou et al. 2017)
PbSe	50-100	Solution	300	125	0.8	0.04	(Liang et al. 2009)
Si	20	SNAP	200	$20 \cdot 10^4$	3.5	0.685	(Boukai et al. 2008)
Si	50 rough	Aq. EE	240	$5.8 \cdot 10^4$	1.6	0.6	(Hochbaum et al. 2008)
Si	20-200	MACE	382	$1.3 \cdot 10^4$	1.68	0.493	(Zhang et al. 2015)
Si	70-130	CVD	210	$2.9 \cdot 10^4$	20	0.02	(Díez et al. 2020)
Si	50-250	MACE	540	$1 \cdot 10^4$	38	0.023	(Xu et al. 2014)
Si	218 pores	MACE	200	$5 \cdot 10^4$	2	0.31	(Yang et al. 2021)
Si_xGe_y	26	VLS	-100	$9 \cdot 10^4$	2	0.13	(Lee et al. 2012)
$CrSi_2$	78	CVT	125	$10 \cdot 10^4$	8	0.06	(Zhou et al. 2007)
$CsSnI_3$	400-800	Solution	80	$2.8 \cdot 10^4$	0.55	0.097	(Lee et al. 2017)
$La_{1-x}Sr_xCoO_3$	83	Hydroth.	213	$3.5\ 10^4$	0.35	0.17	(Wang Fan 2010)

Table 3.1. *Bibliography review of measured properties of different thermoelectric nanowires, where the diameter (Diam, nm), composition (Compos), Seebeck coefficient (S), electrical conductivity (σ), thermal conductivity (κ) and zT at room temperature are collected, as well as the reference*

COMMENTS ON TABLE 3.1.– *Please note that the values shown in this table may not be the highest value shown in the papers, but all the values have been collected at RT for comparison. The growth techniques are the following: Sol-pellets, solution-based fabrication and then SPS sintering; ED-AAO, template-assisted electrochemical deposition in alumina; ED-PC, template-assisted electrochemical deposition in porous polycarbonate; CBE, chemical beam epitaxy; VLS, vapor-liquid-solid; HT, heat treatment; Th-co-ev, thermal co-evaporation; Au-assist chemBE, Au-assisted chemical beam epitaxy; Aq. EE, aqueous electroless etching; CVT, chemical vapor transport; MACE, Metal-Assisted Chemical Etching; SNAP, superlattice nanowire pattern transfer method; PVD, Physical Vapor Deposition; Stress ind., stress-induced growth method; SL, super lattice, periodical composition changes.*

It can be seen that the influence of the dimensionality reduction in materials that are considered as good thermoelectric materials is not so relevant when achieving zT values below or similar to those reported in bulk. However, it can be observed that the smaller the diameter, the higher the zT.

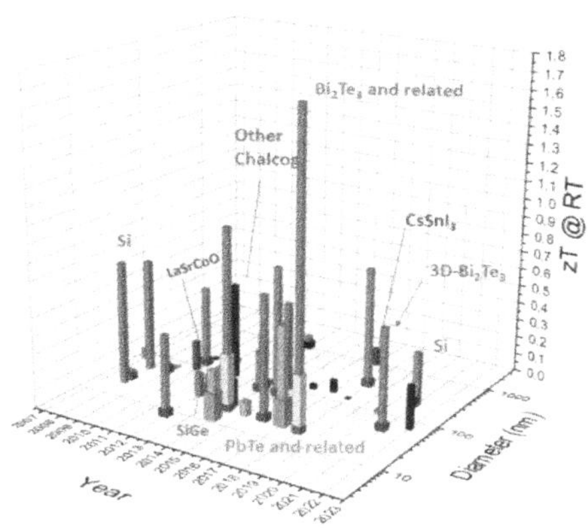

Figure 3.4. *Summary of the reported zT at room temperature (RT) for nanowires of various materials (green for Bi_2Te_3 and related alloys; orange for PbTe and PbSe; blue for other chalcogenides; gray for silicon, SiGe, and $CrSi_2$; and finally red for perovskites and other oxides) based on published values. The zT values at room temperature (RT) are represented versus the nanowire diameters and versus the publication year. The data has been taken for the references listed in Table 3.1. For comparison purposes, the different nanowire values have been compared with the zT of the bulk material at RT, which has been represented along the column values in the cases where they overcome the bulk values for direct comparison. These zT values at RT are 0.01 for bulk Si (Hochbaum et al. 2008), 0.15 for bulk PbTe (Guo et al. 2022), which has been represented in darker orange, and the highest being 0.32 for bulk Bi_2Te_3 (Ruiz-Clavijo et al. 2021) represented in light green. In the case of electrodeposited Bi_2Te_3, it is more accurate to compare the obtained zT of the electrodeposited 3D network and nanowires with that of electrochemically grown Bi_2Te_3 films grown under similar conditions and with the same crystal orientation, since the same technique and conditions are used. This value is of 0.04 at RT (Manzano et al. 2016), which are represented in a darker green*

3.5. Nanowire-based thermoelectric devices

To take advantage of the improved properties of nanowires for thermoelectric applications, along with exploiting the reduced size of NW structures, NW-based thermoelectric generators (TEGs) have been implemented in the last few years. In most cases, a suitable fabrication process, along with the integration of the NWs in the final device, is not straightforward, but the development of micro- and nanotechnology has provided different approaches to doing so.

The most straightforward configuration would be the substitution of the bulk thermoelectric material in a conventional TEG by an array of vertical nanowires (Elyamny et al. 2020; Borshchevsky et al. 2005). Nevertheless, in reality, this configuration needs mechanical stability, which is difficult to achieve in most cases. Dense and aligned NWs are required, and the fabrication of the electrical and thermal contacts is also challenging, although it has been achieved in some cases, even with bottom-up approaches (Dávila et al. 2015). An alternative to increase its robustness is to employ nanowires embedded in a matrix, either the template in which they are grown or dispersed inside a polymeric matrix. In the case of template-assisted growths, the fabrication can be done in such a way that the same matrix is used for both p- and n-type materials and stays in the final TEG (Lim et al. 2005). In the case of dispersing a high number of NWs in a polymer (PEDOT:PSS, for instance), this enables the formation of composites which can even be formulated for printing (Karalis et al. 2021). These kinds of composites are also proposed for fabricating flexible TEGs, which can be used to power wearable devices (see, for instance, Li et al. 2019; Koukharenko et al. 2018). Finally, nanowires pressed into thin films have also been proposed for the design of stretchable TEGs, such as Te, $Cu_{1.75}Te$, or Ag_2Te NWs (Chen et al. 2022).

A different approach is that proposed by Professor Fonseca's group, where CVD-VLS fabricated p- and n-type Si-NWs have been fabricated in a horizontal configuration within a microdevice (Dávila et al. 2012). Apart from being completely compatible with CMOS, this kind of TEGs is improving its performance with modifications in the design, NW fabrication, etc., obtaining powers in the range of $\mu W \cdot cm^{-1}$ (Fonseca et al. 2021, Sojo-Gordillo et al. 2022). A similar approach has been used to develop TEGs based on InAs NWs of diameters around 540 nm, reaching lower produced powers in the range of NW (Koskinen et al. 2021).

3.6. Interconnected 3D nanowire networks

A novel configuration that, on the one hand, overcomes the difficulties for the integration of thermoelectric NWs in bulk devices in a vertical configuration, and on the other hand, maintains the NW properties, has been developed lately: the fabrication of 3D interconnected nanowire networks. These structures provide a mixture between the enhanced thermoelectric properties of low-dimensional structures with the robustness of a scaffold-like structure. Then, these structures can be handled at the macroscale and their implementation in TEGs is much straightforward. These self-sustained structures can be obtained by electrochemical template-assisted growth, provided the template has a 3D structure. This can be performed in polycarbonate (PC) films where a net of nanotubes can be formed by irradiating the PC at two different angles and then etching them. The growth of a thermoelectric material produces an interconnected structure of nanowires (see Figure 3.5a (Wagner et al. 2021)). Also, these PC films have been used to fabricate interconnected NWs of magnetic materials (CoNi, Cu or Permalloy) and even multilayered 3D networks, which show a thermoelectirc voltage which can be further controlled in the presence of a magnetic field (da Câmara Santa Clara Gomes et al. 2022; Marchal et al. 2020). Another way of fabricating 3D templates is the fabrication of 3D Alumina templates (3D-AAOs), proposed in 2014 by Martín-González et al. (2014). These 3D-AAOs consist of longitudinal nanotubes which are interconnected with their first neighbors at certain positions, which can be tailored during the fabrication. Then, the filling of such structures with thermoelectric materials, as Bi_2Te_3 (Ruiz-Clavijo et al. 2018) produces an ordered 3D nanowire network at the macroscopic scale (even after the dissolution of the matrix), which maintains the properties of nanowire configuration (see Figure 3.5b). For instance, the thermal conductivity, obtained for a 3D network of Bi_2Te_3 NWs, is reduced to 0.5 $W{\cdot}m^{-1}{\cdot}K^{-1}$, as compared to 2.2 $W{\cdot}m^{-1}{\cdot}K^{-1}$ for films or to 0.72 $W{\cdot}m^{-1}{\cdot}K^{-1}$ when compared to NW of the same diameters (50 nm), as shown in (Abad et al. 2016). Moreover, recent results from our group have shown that these structures preserve the high electrical conductivity found in electrodeposited Bi_2Te_3 films and an increase in the Seebeck coefficient when compared to electrodeposited nanowires and films has been measured. This latter effect may be caused by the topological surface states that are present in nanowires (given that Bi_2Te_3 is a topological insulator material and the high quality of the 3D nanowires that have been measured) (Ruiz-Clavijo et al. 2021). These results on 3D nanowire networks open a new research avenue to implement nanowires into devices with improved thermoelectric performance versus 1D nanowires or films.

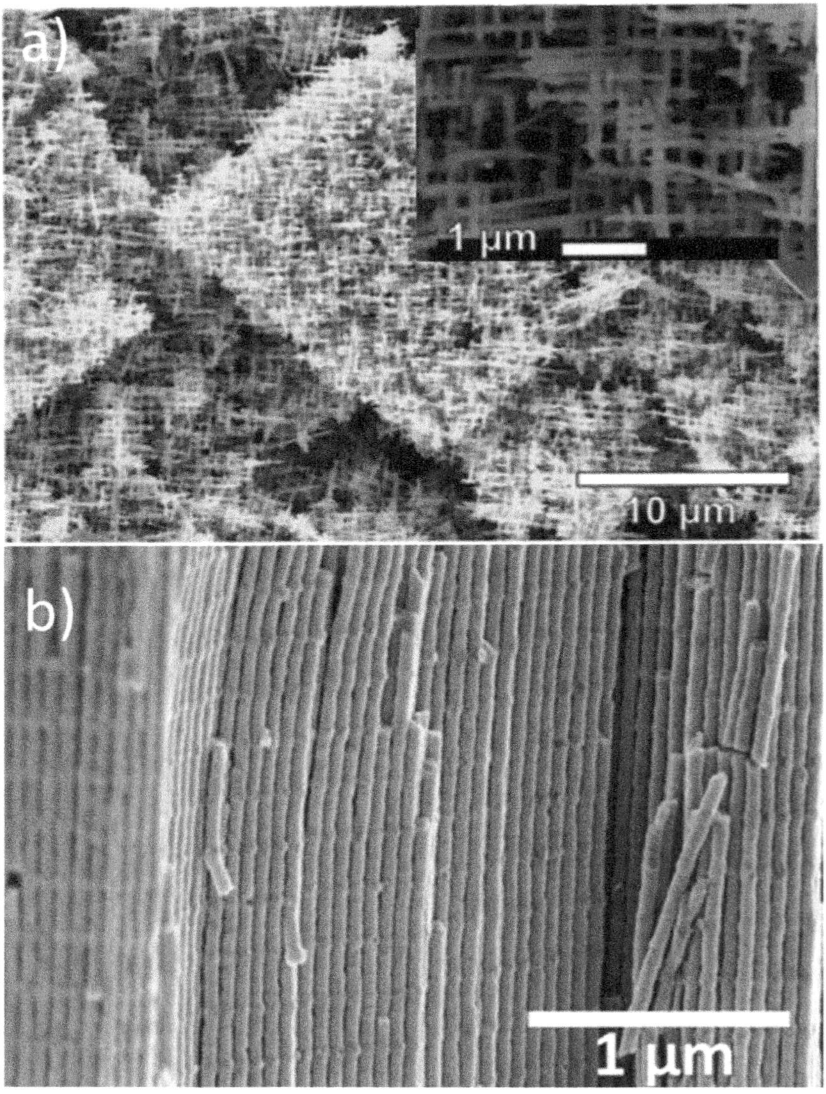

Figure 3.5. *(a) 3D Bi nanowire array grown inside a PC membrane at different magnifications (adapted from (Wagner et al. 2021)). (b) 3D-Bi_2Te_3 nanowire network grown inside a 3D-AAO (the 3D nanowire network is free standing after the membrane is dissolved (courtesy of Dr. A. Ruiz-Clavijo))*

3.7. Summary and outlook

In this chapter, we have shown the latest developments in thermoelectric nanowires, from the motivation behind their development to the different techniques used for their fabrication to their implementation in actual generators, which is a field in continuous development. Moreover, the recently introduced 3D nanowire networks have opened up even more possibilities for developing these structures for thermoelectric applications. This, along with the number of open questions still unanswered, makes the research on thermoelectric nanowires an interesting path to increase the thermoelectric performance of a great number of materials.

The future of this field should be two-fold under our point to view. On the one hand, the study of the underlying physics behind the different parameters and how they can be engineered in nanowires, and in 3D metamaterials, not only to further understand their physics (for instance, how the enhanced the reduction in thermal conductivity or how the surface effect in topological insulators can be controlled to enhance the thermoelecrtric efficiency) but to engineer better-performing devices. On the other hand, the development of nanotechnology tools will enable the fabrication of even more complex and controlled nanowire structures, along with possibilities of more reliable measurement techniques at the nanoscale and better metrology. These ever-growing possibilities should pave the way to novel breakthroughs in the performance of these low dimensional structures and their implementation in efficiency enhanced thermoelectric generators.

3.8. References

Abad, B., Maiz, J., Ruiz-Clavijo, A., Caballero-Calero, O., Martín-González, M. (2016). Tailoring thermal conductivity via three-dimensional porous alumina. *Scientific Reports*, 6(1), 1–10.

Abad, B., Borca-Tasciuc, D.-A., Martín-González, M. (2017). Non-contact methods for thermal properties measurement. *Renewable and Sustainable Energy Reviews*, 76(1), 1348–1370.

Bisrat, Y., Luo, Z., Davis, D., Lagoudas, D. (2007). Highly ordered uniform single-crystal Bi nanowires: Fabrication and characterization. *Nanotechnology*, 18(39), 395601.

Biswas, K.G., Sands, T.D., Cola, B.A., Xu, X. (2009). Thermal conductivity of bismuth telluride nanowire array-epoxy composite. *Applied Physics Letters*, 94(22), 223116.

Borshchevsky, A., Fleurial, J.-P., Herman, J., Ryan, M. (2005). Nanowire thermoelectric devices. Publication NASA Tech Briefs, June 2005, NPO – 30210. Document ID: 20110014928.

Boukai, A.I., Bunimovich, Y., Tahir-Kheli, J., Yu, J.-K., Goddard III, W.A., Heath, J.R. (2008). Silicon nanowires as efficient thermoelectric materials. *Nature*, 451(7175), 168–171.

Brockway, L., Vasiraju, V., Asayesh-Ardakani, H., Shahbazian-Yassar, R., Vaddiraju, S. (2014). Thermoelectric properties of large-scale Zn3P2 nanowire assemblies. *Nanotechnology*, 25(14), 145401.

Cassinelli, M., Müller, S., Voss, K.-O., Trautmann, C., Völklein, F., Gooth, J., Nielsch, K., Toimil-Molares, M. (2017). Influence of surface states and size effects on the Seebeck coefficient and electrical resistance of $Bi_{1-x}Sb_x$ nanowire arrays. *Nanoscale*, 9(9), 3169–3179.

Chen, C.-L., Chen, Y.-Y., Lin, S.-J., Ho, J.C., Lee, P.-C., Chen, C.-D., Harutyunyan, S.R. (2010). Fabrication and characterization of electrodeposited bismuth telluride films and nanowires. *The Journal of Physical Chemistry C*, 114(8), 3385–3389.

Chen, C., Wang, R., Li, X.-L., Zhao, B., Wang, H., Zhou, Z., Zhu, J., Liu, J.-W. (2022). Structural design of nanowire wearable stretchable thermoelectric generator. *Nano Letters*, 10, 4131–4136.

Da Câmara Santa Clara Gomes, T., Marchal, N., Abreu Araujo, F., Piraux, L. (2022). Magnetically activated flexible thermoelectric switches based on interconnected nanowire networks. *Advanced Materials Technologies*, 7(5), 2101043.

Dávila, D., Tarancón, A., Calaza, C., Salleras, M., Fernández-Regúlez, M., San Paulo, A., Fonseca, L. (2012). Monolithically integrated thermoelectric energy harvester based on silicon nanowire arrays for powering micro/nanodevices. *Nano Energy*, 1(6), 812–819.

Dávila, D., Huber, R., Hierold, C. (eds) (2015). Bottom-up silicon nanowire-based thermoelectric microgenerators. *Journal of Physics: Conference Series*, IOP Publishing, 012101.

Díez, G.G., Gordillo, J.M.S., Pujadó, M.P., Salleras, M., Fonseca, L., Morata, A., Rubio, A.T. (2020). Enhanced thermoelectric figure of merit of individual Si nanowires with ultralow contact resistances. *Nano Energy*, 67(1), 104191.

Ding, C., Lu, T., Wazir, N., Ma, W., Guo, S., Xin, Y., Li, A., Liu, R., Zou, B. (2021). New type of thermoelectric CdSSe nanowire chip. *ACS Applied Materials & Interfaces*, 13(26), 30959–30966.

Elyamny, S., Dimaggio, E., Magagna, S., Narducci, D., Pennelli, G. (2020). High power thermoelectric generator based on vertical silicon nanowires. *Nano Letters*, 20(7), 4748–4753.

Fang, H., Feng, T., Yang, H., Ruan, X., Wu, Y. (2013). Synthesis and thermoelectric properties of compositional-modulated lead telluride–bismuth telluride nanowire heterostructures. *Nano Letters*, 13(5), 2058–2063.

Fang, H., Yang, H., Wu, Y. (2014). Thermoelectric properties of silver telluride–bismuth telluride nanowire heterostructure synthesized by site-selective conversion. *Chemistry of Materials*, 26(10), 3322–3327.

Finefrock, S.W., Zhang, G., Bahk, J.-H., Fang, H., Yang, H., Shakouri, A., Wu, Y. (2014). Structure and thermoelectric properties of spark plasma sintered ultrathin PbTe nanowires. *Nano Letters*, 14(6), 3466–3473.

Fonseca, L., Donmez-Noyan, I., Dolcet, M., Estrada-Wiese, D., Santander, J., Salleras, M., Gadea, G., Pacios, M., Sojo, J.-M., Morata, A. (2021). Transitioning from Si to SiGe nanowires as thermoelectric material in silicon-based microgenerators. *Nanomaterials*, 11(2), 517.

Fust, S., Faustmann, A., Carrad, D.J., Bissinger, J., Loitsch, B., Döblinger, M., Becker, J., Abstreiter, G., Finley, J.J., Koblmüller, G. (2020). Quantum-confinement-enhanced thermoelectric properties in modulation-doped GaAs–AlGaAs core–shell nanowires. *Advanced Materials*, 32(4), 1905458.

Ghukasyan, A., Oliveira, P., Goktas, N.I., Lapierre, R. (2022). Thermal conductivity of GaAs nanowire arrays measured by the 3ω method. *Nanomaterials*, 12(8), 1288.

Gorokh, G., Lozovenko, A., Bulat, L. (2017). Thermoelectric battery based on bundles of Bi and Sb nanowires in anodic alumina matrices. *Semiconductors*, 51(7), 850–853.

Guo, D., Zhang, H., Xu, Z., Li, C., Li, K., Shao, B., Chen, D., Ma, Y., Sun, J. (2022). The room-temperature thermoelectric property of PbTe enhanced by mean-free-path filtering. *Journal of Alloys and Compounds*, 893(162296).

Hamdana, G., Südkamp, T., Descoins, M., Mangelinck, D., Caccamo, L., Bertke, M., Wasisto, H.S., Bracht, H., Peiner, E. (2017). Towards fabrication of 3D isotopically modulated vertical silicon nanowires in selective areas by nanosphere lithography. *Microelectronic Engineering*, 179(1), 74–82.

Hernandez, J.A., Ruiz, A., Fonseca, L.F., Pettes, M.T., Jose-Yacaman, M., Benitez, A. (2018). Thermoelectric properties of SnSe nanowires with different diameters. *Scientific Reports*, 8(1), 1–8.

Hicks, L.D. and Dresselhaus, M.S. (1993). Thermoelectric figure of merit of a one-dimensional conductor. *Physical Review B*, 47(24), 16631.

Hochbaum, A.I., Chen, R., Delgado, R.D., Liang, W., Garnett, E.C., Najarian, M., Majumdar, A., Yang, P. (2008). Enhanced thermoelectric performance of rough silicon nanowires. *Nature*, 451(7175), 163–167.

Hsin, C.-L., Wingert, M., Huang, C.-W., Guo, H., Shih, T.-J., Suh, J., Wang, K., Wu, J., Wu, W.-W., Chen, R. (2013). Phase transformation and thermoelectric properties of bismuth-telluride nanowires. *Nanoscale*, 5(11), 4669–4672.

Karalis, G., Tzounis, L., Mytafides, C.K., Tsirka, K., Formanek, P., Stylianakis, M., Kymakis, E., Paipetis, A.S. (2021). A high performance flexible and robust printed thermoelectric generator based on hybridized Te nanowires with PEDOT:PSS. *Applied Energy*, 294(1), 117004.

Kim, J., Oh, M.-W., Kim, G., Bahk, J.-H., Song, J.Y., Jeon, S.G., Chun, D.W., Bae, J.-H., Shim, W., Lee, W. (2018). Strain-engineered allotrope-like bismuth nanowires for enhanced thermoelectric performance. *Acta Materialia*, 144(1), 145–153.

Kim, M., Park, D., Kim, J. (2021). Synergistically enhanced thermoelectric performance by optimizing the composite ratio between hydrothermal Sb_2Se_3 and self-assembled β-Cu_2Se nanowires. *CrystEngComm*, 23(15), 2880–2888.

Koga, T., Sun, X., Cronin, S., Dresselhaus, M. (1998). Carrier pocket engineering to design superior thermoelectric materials using GaAs/AlAs superlattices. *Applied Physics Letters*, 73(20), 2950–2952.

Kong, D., Randel, J.C., Peng, H., Cha, J.J., Meister, S., Lai, K., Chen, Y., Shen, Z.-X., Manoharan, H.C., Cui, Y. (2010). Topological insulator nanowires and nanoribbons. *Nano Letters*, 10(1), 329–333.

Koskinen, T., Khayrudinov, V., Emadi, F., Jiang, H., Haggren, T., Lipsanen, H., Tittonen, I. (2021). Thermoelectric characteristics of InAs nanowire networks directly grown on flexible plastic substrates. *ACS Applied Energy Materials*, 4(12), 14727–14734.

Koukharenko, E., Boden, S., Sessions, N., Frety, N., Nandhakumar, I., White, N. (2018). Towards thermoelectric nanostructured energy harvester for wearable applications. *Journal of Materials Science: Materials in Electronics*, 29(4), 3423–3436.

Lange, F., Ernst, O., Teubner, T., Richter, C., Schmidbauer, M., Skibitzki, O., Schroeder, T., Schmidt, P., Boeck, T. (2020). In-plane growth of germanium nanowires on nanostructured $Si_{(001)}/SiO_2$ substrates. *Nano Futures*, 4(3), 035006.

Lee, S.H., Shim, W., Jang, S.Y., Roh, J.W., Kim, P., Park, J., Lee, W. (2011). Thermoelectric properties of individual single-crystalline PbTe nanowires grown by a vapor transport method. *Nanotechnology*, 22(29), 295707.

Lee, E.K., Yin, L., Lee, Y., Lee, J.W., Lee, S.J., Lee, J., Cha, S.N., Whang, D., Hwang, G.S., Hippalgaonkar, K. et al. (2012). Large thermoelectric figure-of-merits from SiGe nanowires by simultaneously measuring electrical and thermal transport properties. *Nano Letters*, 12(6), 2918–2923.

Lee, W., Li, H., Wong, A.B., Zhang, D., Lai, M., Yu, Y., Kong, Q., Lin, E., Urban, J.J., Grossman, J.C. (2017). Ultralow thermal conductivity in all-inorganic halide perovskites. *Proceedings of the National Academy of Sciences*, 114(33), 8693–8697.

Li, L., Xu, S., Li, G. (2015). Enhancement of thermoelectric properties in Bi–Sb–Te alloy nanowires by pulsed electrodeposition. *Energy Technology*, 3(8), 825–829.

Li, Z., Xu, E., Losovyj, Y., Li, N., Chen, A., Swartzentruber, B., Sinitsyn, N., Yoo, J., Jia, Q., Zhang, S. (2017). Surface oxidation and thermoelectric properties of indium-doped tin telluride nanowires. *Nanoscale*, 9(35), 13014–13024.

Li, C., Jiang, F., Liu, C., Liu, P., Xu, J. (2019). Present and future thermoelectric materials toward wearable energy harvesting. *Applied Materials Today*, 15(1), 543–557.

Liang, W., Rabin, O., Hochbaum, A.I., Fardy, M., Zhang, M., Yang, P. (2009). Thermoelectric properties of p-type PbSe nanowires. *Nano Research*, 2(5), 394–399.

Lim, J.R., Whitacre, J.F., Fleurial, J.P., Huang, C.K., Ryan, M.A., Myung, N.V. (2005). Fabrication method for thermoelectric nanodevices. *Advanced Materials*, 17(12), 1488–1492.

Llobet, J., Rius, G., Chuquitarqui, A., Borrisé, X., Koops, R., Van Veghel, M., Perez-Murano, F. (2018). Arrays of suspended silicon nanowires defined by ion beam implantation: Mechanical coupling and combination with CMOS technology. *Nanotechnology*, 29(15), 155303.

Mallavarapu, A., Ajay, P., Barrera, C., Sreenivasan, S. (2020). Ruthenium-assisted chemical etching of silicon: Enabling CMOS-compatible 3D semiconductor device nanofabrication. *ACS Applied Materials & Interfaces*, 13(1), 1169–1177.

Manzano, C.V. and Martín-González, M. (2019). Electrodeposition of V-VI nanowires and their thermoelectric properties. *Frontiers in Chemistry*, 7(1), 516.

Manzano, C.V., Polyakov, M.N., Maiz, J., Aguirre, M.H., Maeder, X., Martín-González, M. (2019). Pulsed current-voltage electrodeposition of stoichiometric Bi_2Te_3 nanowires and their crystallographic characterization by transmission electron backscatter diffraction. *Science and Technology of Advanced Materials*, 20(1), 1022–1030.

Marchal, N., Da Câmara Santa Clara Gomes, T., Abreu Araujo, F., Piraux, L. (2020). Large spin-dependent thermoelectric effects in NiFe-based interconnected nanowire networks. *Nanoscale Research Letters*, 15(1), 1–8.

Martín, J., Martín-González, M., Francisco Fernández, J., Caballero-Calero, O. (2014). Ordered three-dimensional interconnected nanoarchitectures in anodic porous alumina. *Nature Communications*, 5(1), 5130.

Martinez, J., Martinez, R.V., Garcia, R. (2008). Silicon nanowire transistors with a channel width of 4 nm fabricated by atomic force microscope nanolithography. *Nano letters*, 8(11), 3636–3639.

Mavrokefalos, A., Moore, A.L., Pettes, M.T., Shi, L., Wang, W., Li, X. (2009). Thermoelectric and structural characterizations of individual electrodeposited bismuth telluride nanowires. *Journal of Applied Physics*, 105(10), 104318.

Park, D., Park, S., Jeong, K., Jeong, H.-S., Song, J.Y., Cho, M.H. (2016). Thermal and electrical conduction of single-crystal Bi_2Te_3 nanostructures grown using a one step process. *Scientific Reports*, 6(1), 1–9.

Pennelli, G. and Piotto, M. (2006). Fabrication and characterization of silicon nanowires with triangular cross section. *Journal of Applied Physics*, 100(5), 054507.

Peri, L., Prete, D., Demontis, V., Zannier, V., Rossi, F., Sorba, L., Beltram, F., Rossella, F. (2022). Giant reduction of thermal conductivity and enhancement of thermoelectric performance in twinning superlattice InAsSb nanowires. *Nano Energy*, 103(107700).

Pillai, K. (2021). Growth of single-crystalline highly pure long Ag_2Te nanowires for thermoelectric applications and its structural analysis. *Journal of Materials Science: Materials in Electronics*, 1–8.

Qiu, X., Lou, Y., Samia, A.C., Devadoss, A., Burgess, J.D., Dayal, S., Burda, C. (2005). PbTe nanorods by sonoelectrochemistry. *Angewandte Chemie International Edition*, 44(36), 5855–5857.

Rojo, M.M., Calero, O.C., Lopeandia, A., Rodriguez-Viejo, J., Martín-González, M. (2013a). Review on measurement techniques of transport properties of nanowires. *Nanoscale*, 5(23), 11526–11544.

Rojo, M.M., Grauby, S., Rampnoux, J.-M., Caballero-Calero, O., Martín-González, M., Dilhaire, S. (2013b). Fabrication of Bi_2Te_3 nanowire arrays and thermal conductivity measurement by 3ω-scanning thermal microscopy. *Journal of Applied Physics*, 113(5), 054308.

Rojo, M.M., Martín, J., Grauby, S., Borca-Tasciuc, T., Dilhaire, S., Martín-González, M. (2014). Decrease in thermal conductivity in polymeric P_3HT nanowires by size-reduction induced by crystal orientation: New approaches towards thermal transport engineering of organic materials. *Nanoscale*, 6(14), 7858–7865.

Rojo, M.M., Zhang, Y., Manzano, C.V., Alvaro, R., Gooth, J., Salmeron, M., Martín-González, M. (2016). Spatial potential ripples of azimuthal surface modes in topological insulator Bi_2Te_3 nanowires. *Scientific Reports*, 6(1), 1–8.

Rojo, M.M., Abad, B., Manzano, C., Torres, P., Cartoixà, X., Alvarez, F., Gonzalez, M.M. (2017). Thermal conductivity of Bi_2Te_3 nanowires: How size affects phonon scattering. *Nanoscale*, 9(20), 6741–6747.

Ruiz-Clavijo, A., Caballero-Calero, O., Martín-González, M. (2018). Three-dimensional Bi_2Te_3 networks of interconnected nanowires: Synthesis and optimization. *Nanomaterials*, 8(5), 345.

Ruiz-Clavijo, A., Caballero-Calero, O., Manzano, C.V., Maeder, X., Beardo, A., Cartoixà, X., Álvarez, F.X., Martín-González, M. (2021). 3D Bi_2Te_3 interconnected nanowire networks to increase thermoelectric efficiency. *ACS Applied Energy Materials*, 4, 13556–13566.

Ruoho, M., Juntunen, T., Tittonen, I. (2016). Large-area thermoelectric high-aspect-ratio nanostructures by atomic layer deposition. *Nanotechnology*, 27(35), 355403.

Shin, H.S., Jeon, S.G., Yu, J., Kim, Y.-S., Park, H.M., Song, J.Y. (2014). Twin-driven thermoelectric figure-of-merit enhancement of Bi_2Te_3 nanowires. *Nanoscale*, 6(11), 6158–6165.

Sojo-Gordillo, J.M., Diez, G.G., Pujadó, M.P., Salleras, M., Estrada-Wiese, D., Dolcet, M., Fonseca, L., Morata, A., Tarancón, A. (2021). Thermal conductivity of individual Si and SiGe epitaxially integrated NWs by scanning thermal microscopy. *Nanoscale*, 13(15), 7252–7265.

Sojo-Gordillo, J.M., Estrada-Wiese, D., Duque-Sierra, C., Gadea-Díez, G., Salleras, M., Fonseca, L., Morata, A., Tarancón, A. (2022). Tuning the thermoelectric properties of boron-doped silicon nanowires integrated into a micro-harvester. *Advanced Materials Technologies*, 2101715.

Sun, K.W., Ko, T.-Y., Shellaiah, M. (2018). Field-effect-dependent thermoelectric power in highly resistive Sb_2Se_3 single nanowire. *Applied Physics A*, 124(4), 1–8.

Tai, G.A., Zhou, B., Guo, W. (2008). Structural characterization and thermoelectric transport properties of uniform single-crystalline lead telluride nanowires. *The Journal of Physical Chemistry C*, 112(30), 11314–11318.

Tan, M., Deng, Y., Wang, Y. (2014). Ordered structure and high thermoelectric properties of Bi_2 (Te, Se)$_3$ nanowire array. *Nano Energy*, 3(1), 144–151.

Tan, M., Hao, Y., Deng, Y., Chen, J. (2018). High thermoelectric properties of (Sb, Bi)$_2$Te$_3$ nanowire arrays by tilt-structure engineering. *Applied Surface Science*, 443(1), 11–17.

Valentín, L., Betancourt, J., Fonseca, L., Pettes, M., Shi, L., Soszyński, M., Huczko, A. (2013). A comprehensive study of thermoelectric and transport properties of β-silicon carbide nanowires. *Journal of Applied Physics*, 114(18), 184301.

Vivekanandan, A.K., Lee, C.-W., Wu, R.-Z., Tsai, W.-H., Chen, S.-H., Chen, Y.-Y., Lin, C.-T. (2022). Tailoring InSb nanowires for high thermoelectric performance using AAO template-assisted die casting process. *Nanomaterials*, 12(12), 2032.

Wagner, A.R. and Ellis, S.W. (1964). Vapor-liquid-solid mechanism of single crystal growth. *Applied Physics Letters*, 4(5), 89–90.

Wagner, M.F.P., Paulus, A.S., Brötz, J., Sigle, W., Trautmann, C., Voss, K.O., Völklein, F., Toimil-Molares, M.E. (2021). Effects of size reduction on the electrical transport properties of 3D Bi nanowire networks. *Advanced Electronic Materials*, 7(3), 2001069.

Wang, Y. and Fan, H.J. (2010). Improved thermoelectric properties of La_{1-x}Srx CoO3 nanowires. *The Journal of Physical Chemistry C*, 114(32), 13947–13953.

Wang, G., Lok, S., Wong, G., Sou, I. (2009). Molecular beam epitaxy-grown Bi_4Te_3 nanowires. *Applied Physics Letters*, 95(26), 263102.

Wu, P.M., Gooth, J., Zianni, X., Svensson, S.F., Gluschke, J.G.R., Dick, K.A., Thelander, C., Nielsch, K., Linke, H. (2013). Large thermoelectric power factor enhancement observed in InAs nanowires. *Nano letters*, 13(9), 4080–4086.

Xu, B., Khouri, W., Fobelets, K. (2014). Two-sided silicon nanowire array/bulk thermoelectric power generator. *IEEE Electron Device Letters*, 35(5), 596–598.

Yang, R.B., Zakharov, N., Moutanabbir, O., Scheerschmidt, K., Wu, L.-M., GöSele, U., Bachmann, J., Nielsch, K. (2010). The transition between conformal atomic layer epitaxy and nanowire growth. *Journal of the American Chemical Society*, 132(22), 7592–7594.

Yang, H., Bahk, J.-H., Day, T., Mohammed, A.M., Snyder, G.J., Shakouri, A., Wu, Y. (2015). Enhanced thermoelectric properties in bulk nanowire heterostructure-based nanocomposites through minority carrier blocking. *Nano Letters*, 15(2), 1349–1355.

Yang, L., Huh, D., Ning, R., Rapp, V., Zeng, Y., Liu, Y., Ju, S., Tao, Y., Jiang, Y., Beak, J. (2021). High thermoelectric figure of merit of porous Si nanowires from 300 to 700 K. *Nature Communications*, 12(1), 1–7.

Yang, X., Wang, C., Lu, R., Shen, Y., Zhao, H., Li, J., Li, R., Zhang, L., Chen, H., Zhang, T. (2022). Progress in measurement of thermoelectric properties of micro/nano thermoelectric materials: A critical review. *Nano Energy*, 107553.

Yazji, S., Swinkels, M.Y., De Luca, M., Hoffmann, E., Ercolani, D., Roddaro, S., Abstreiter, G., Sorba, L., Bakkers, E.P., Zardo, I. (2016). Assessing the thermoelectric properties of single InSb nanowires: The role of thermal contact resistance. *Semiconductor Science and Technology*, 31(6), 064001.

Zhang, Z., Gekhtman, D., Dresselhaus, M.S., Ying, J.Y. (1999). Processing and characterization of single-crystalline ultrafine bismuth nanowires. *Chemistry of Materials*, 11(7), 1659–1665.

Zhang, G., Wang, W., Li, X. (2008). Enhanced thermoelectric properties of core/shell heterostructure nanowire composites. *Advanced Materials*, 20(19), 3654–3656.

Zhang, G., Fang, H., Yang, H., Jauregui, L.A., Chen, Y.P., Wu, Y. (2012a). Design principle of telluride-based nanowire heterostructures for potential thermoelectric applications. *Nano Letters*, 12(7), 3627–3633.

Zhang, G., Kirk, B., Jauregui, L.A., Yang, H., Xu, X., Chen, Y.P., Wu, Y. (2012b). Rational synthesis of ultrathin n-type Bi_2Te_3 nanowires with enhanced thermoelectric properties. *Nano Letters*, 12(1), 56–60.

Zhang, T., Wu, S., Xu, J., Zheng, R., Cheng, G. (2015). High thermoelectric figure-of-merits from large-area porous silicon nanowire arrays. *Nano Energy*, 13(1), 433–441.

Zhou, F., Szczech, J., Pettes, M.T., Moore, A.L., Jin, S., Shi, L. (2007). Determination of transport properties in chromium disilicide nanowires via combined thermoelectric and structural characterizations. *Nano Letters*, 7(6), 1649–1654.

Zhou, F., Moore, A.L., Pettes, M.T., Lee, Y., Seol, J.H., Ye, Q.L., Rabenberg, L., Shi, L. (2009). Effect of growth base pressure on the thermoelectric properties of indium antimonide nanowires. *Journal of Physics D: Applied Physics*, 43(2), 025406.

Zhou, C., Dun, C., Wang, K., Zhang, X., Shi, Z., Liu, G., Hewitt, C.A., Qiao, G., Carroll, D.L. (2016). General method of synthesis ultrathin ternary metal chalcogenide nanowires for potential thermoelectric applications. *Nano Energy*, 30(1), 709–716.

Zhou, C., Shi, Z., Ge, B., Wang, K., Zhang, D., Liu, G., Qiao, G. (2017). Scalable solution-based synthesis of component-controllable ultrathin $PbTe_{1-x}Se_x$ nanowires with high n-type thermoelectric performance. *Journal of Materials Chemistry A*, 5(6), 2876–2884.

Zhu, Q., Wang, S., Wang, X., Suwardi, A., Chua, M.H., Soo, X.Y.D., Xu, J. (2021). Bottom-up engineering strategies for high-performance thermoelectric materials. *Nano-Micro Letters*, 13(1), 1–38.

4

Impact of Chemical Doping or Magnetism in Model Thermoelectric Sulfides

Sylvie HÉBERT, Ramzy DAOU and Antoine MAIGNAN
ENSICAEN, UNICAEN, CNRS, CRISMAT, Normandie Université, Caen, France

The optimization of the thermoelectric properties of chalcogenides is classically performed by modifying the doping and the micro- or nanostructure, by using materials with complex crystallographic structures or by optimizing the defects chemistry. Among all these possibilities, we will highlight the results obtained for model systems among sulfides. First, the investigation of doped TiS_2 single crystals and dense textured ceramics will be presented, along with the determination of doping and anisotropy effects in these 2D structural types. Second, in transition metal chalcogenides, the magnetism can strongly modify transport, with a direct influence on the thermoelectric properties, as shown in the case of itinerant (CoS_2 and Co_9S_8) or localized ($CuCrTiS_4$) carriers in denser 3D structures. These two examples picked up among sulfides illustrate the richness of chalcogenides for thermoelectricity.

For a color version of all the figures in this chapter, see www.iste.co.uk/akinaga/thermoelectric2.zip.

Thermoelectric Micro/Nano Generators 2,
coordinated by Hiroyuki AKINAGA, Atsuko KOSUGA,
Takao MORI and Gustavo ARDILA.

4.1. Introduction

To overcome the limitations of Bi_2Te_3-based thermoelectric (TE) materials related to their decomposition at low T, the tellurium toxicity and scarcity, great attention was given to transition metal oxides offering much larger chemical stability at high T (Hébert et al. 2001; Guilmeau et al. 2009; Koumoto et al. 2010). Although promising TE power factors ($PF = S^2/\rho$ with S and ρ for the Seebeck coefficient and electrical resistivity, respectively) were achieved, as, for instance, in the so-called misfit cobaltites (Sotelo et al. 2011), their too large thermal conductivity, for a textured $Ca_3Co_4O_9$ dense ceramic $\kappa_{300K} = 4.4\ W.K^{-1}.m^{-1}$ (Torres et al. 2019), led to too low values for their figure of merit ($zT = S^2T/(\rho\kappa) = (PF)T/\kappa$; κ is the sum of the electronic (κ_{el}) and lattice (κ_l) parts). Nonetheless, such 2D materials, where the Co cations sit at the center of octahedra forming CdI_2-type metallic layers separated by rock-salt type insulating layers (Masset et al. 2000), could be considered as a realization of the PGEC Slack concept, i.e. phonon glass electron crystal (Slack 1997), especially if we consider the misfit structural character of the sublattices, CdI_2- and rock salt-types.

In that respect, layered transition metal disulfides with similar networks of edge-shared MS_6 octahedra were rapidly proposed as an alternative to CoO_2-based 2D magnetic oxides (Imai et al. 2001; Wan et al. 2010; Guilmeau et al. 2011; Hébert et al. 2013a; Guilmeau et al. 2015). Among them, TiS_2 is close to being a textbook case (1T common form, *P-3m1* hexagonal layered structure): its $[TiS_2]_\infty$ successive layers are separated by a van der Waals (VDW) gap, allowing chemical substitutions in the former and/or cation intercalation (including self-intercalation) in the latter to be made (Guilmeau et al. 2015). The consequences of adding cations in the VDW gap and/or substituting aliovalent cations at the Ti^{4+} site are double, change in (Ti^{3+}) electron concentration (band filling) and hindering of phonon propagation (κ_l decrease).

Another route has been offered by chalcogenides containing transition metal cations with partially filled d orbitals belonging to strongly correlated systems: their TE properties are sensitive to the interplay between charges and spins (Maignan et al. 2003; Wang et al. 2003; Behnia et al. 2004), as probed by magnetic field-dependent thermoelectric power (MTEP) (Hébert et al. 2021a). As the Seebeck coefficient depends on both transport coefficients and entropy (Peterson et al. 2007), large orbital or spin degrees of freedom of the carriers result in large S (Koshibae et al. 2001). At first, considering metal transition oxides, the MTEP measurement of Na_xCoO_2

(Wang et al. 2003) and misfit cobaltites (Maignan et al. 2003) demonstrated this spin entropy role. More recently, a similar role of spins was evidenced for the spin glass $CuCrTiS_4$ thiospinel, for which MTEP was reported (Berthebaud et al. 2018).

In the present chapter, we report on these two different strategies by focusing, for the first part, on TiS_2-based TE materials, and for the second, on two Co-S binary sulfides, Co_9S_8 pentlandite (Hébert et al. 2021b) and CoS_2 pyrite (Hébert et al. 2013b), and on the $CuCrTiS_4$ quaternary thiospinel (Berthebaud et al. 2018; Hébert et al. 2021a). Such examples give only a very partial overview of the numerous studies on TE sulfides, such as the very recent ones on compounds derived from minerals, which zT values approach or are close to unity for the p-type side, including tetrahedrite, colusite and copper-deficient binaries $Cu_{2-x}S$ (Powell 2019 and references therein), or, for the n-type ones, $CuFeS_2$ pyrite (Li et al. 2013), $Co_3Sn_2S_3$ shandite (Mangelis et al. 2020), compounds with zT lower than those of their p-type counterparts.

4.2. TiS_2: intercalation chemistry to combine power factor optimization and lattice thermal conductivity degradation

The question of whether TiS_2 is intrinsically semiconducting or semi-metallic (Stoliaroff et al. 2019) comes from the property discrepancy in different publications for compounds of nominal TiS_2 composition. The careful study of SPS-densified ceramics of $Ti_{1+x}S_2$, where x is the Ti excess in the nominal composition to create Ti self-intercalation in the VDW gap, allowed us to give an answer to that long-term problem (Beaumale et al. 2014). $|S_{300K}|$ values as high as 300 $\mu V.K^{-1}$ could be achieved by tuning x precisely, indicating that when un-doped, TiS_2 is rather semiconducting. Also, instead of Ti self-intercalation, other cations can be intercalated on purpose to optimize the TE properties, such as Cu (Guilmeau et al. 2011), Co (Guélou et al. 2016), Fe (Pawula et al. 2019) or Ag (Barbier et al. 2015), but with staging in the latter case, with Ag not intercalated in every VDW gap. Remarkably, the Cu_xTiS_2-densified ceramics with $x > 0$ were found to exhibit larger zT values than the $x = 0$ pristine TiS_2 prepared in similar synthetic conditions, such as for $Cu_{0.05}TiS_2$, where $zT = 0.46$ at 800 K (Guilmeau et al. 2017). Along that intercalation, as Cu brings positive charges, the charge carrier concentration increases (Ti^{3+} electrons), and so ρ and $|S|$ values decrease concomitantly, leading to a rather constant PF. The more important improvement comes from the κ_l decrease outperforming the

κ_e increase, resulting from the ρ decrease evidenced in these SPS-densified ceramics.

As this compound possesses a 2D structure, its anisotropic TE properties are difficult to measure on these platelet-like crystals. Measuring κ in thin crystals of intercalated TiS_2 (grown by iodine vapor transport (Kawano et al. 1989), exhibiting typical thicknesses of 20 μm) can hardly be realized by diffusivity measurements, as laser flash requires a 1 mm thick sample at minimum.

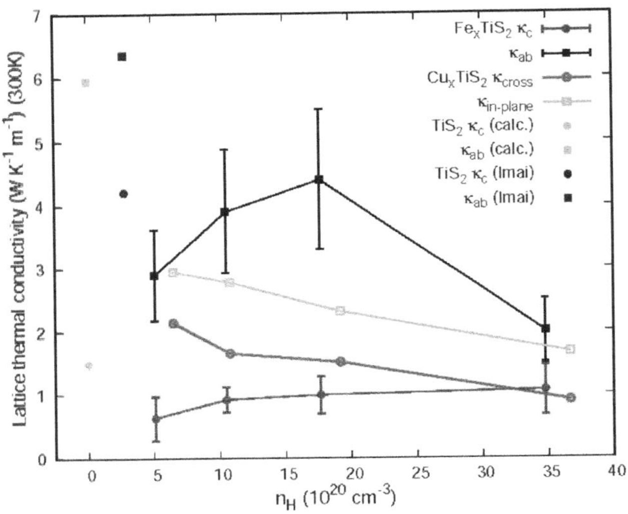

Figure 4.1. *Anisotropy in the lattice thermal conductivity of TiS_2 and derivatives as a function of carrier concentration. Single crystals (Fe_xTiS_2 and TiS_2) and oriented ceramics (Cu_xTiS_2) are compared to the calculated thermal conductivities of the perfect structure derived from ab initio calculations (see the text)*

Using cleaved Fe_xTiS_2 crystals, cross-plane thermal conductivity was measured by picosecond time-domain thermos-reflectance, whereas the in-plane values were measured using the steady-state technique involving a calibrated heat pipe to measure the thermal power entering the sample (Pawula et al. 2019). This allowed us to measure the thermal conductivity anisotropy (Figure 4.1) at room temperature with reasonable uncertainty.

κ_c is found to increase as the Fe content (x) increases, suggesting that intercalated Fe atoms may start to function as phonon conduction paths, rather than as phonon-scattering sources. Moreover, these values, all in the range from $\kappa_c = 0.6$ W.K^{-1}.m^{-1} to $\kappa_c = 1.1$ W.K^{-1}.m^{-1} (Pawula et al. 2019), much smaller than the unique published value for a TiS_2 crystal, κ_c = 4.2 W.K^{-1}.m^{-1} (Imai et al. 2001), point towards a larger κ_{ab}/κ_c anisotropy, of about 6, than the previously reported one, $\kappa_{ab}/\kappa_c \approx 1.5$ (Imai et al. 2001). While a degree of κ anisotropy is also found in textured ceramics (Figure 4.1) (Guilmeau et al. 2017), it is not as marked as in single crystals, which is to be expected given the spread of orientations present. In any case, in comparison to the ρ anisotropy of such 2D materials, of about $\rho_c/\rho_{ab} \approx 750$ in crystals at 300 K (Imai et al. 2001), the high in-plane PF of TiS_2, PF = 37.1ψ µW.K^{-2}.cm (Imai et al. 2001), similar to that of the best Bi_2Te_3 values (Scherrer and Scherrer 1995), is unlikely to be beaten by the c direction, as a strong anisotropy in thermopower is not expected in this material; unlike, for example, similarly structured covalently bonded layered oxide $PdCoO_2$, where the out-of-plane thermoelectric power is predicted to be very large as a result of the strong energy dependence of the conductivity (Ong et al. 2010).

Comparing different crystals, self-intercalated, intercalated by Co, or substituted by Ta for Ti to bring even more electrons (Daou et al. 2015), it turns out that the most efficient way to reduce κ is the substitution, with, at T=300K, κ_{total} = 1.2 W.K^{-1}.m^{-1} in $Ti_{0.9}Ta_{0.1}S_2$ crystal against κ_{total} = 4.3 W.K^{-1}.m^{-1} for our self-intercalated TiS_2 crystal (Daou et al. 2015). But for that $Ti_{1-x}Ta_xS_2$ series, with the charge carrier concentration becoming too large (for x = 0.1, n_e(300 K) = 3.7x10^{21} cm^{-3}), PF drops to about PF = 3.4 µW.K^{-2}.cm, i.e. one order of magnitude smaller than the aforementioned best TiS_2 crystals (Imai et al. 2001). Consequently, this comparison shows that optimizing zT of TiS_2 crystals remains challenging, as the κ_l decrease induced by intercalants or substituents, more efficient in-plane than out-of-plane, is not sufficient to compensate the PF decrease induced by the n_e increase.

Such a conclusion is consistent with those drawn from the study of in-plane and out-of-plane TE measurements in Cu_xTiS_2 SPS prepared textured ceramics (Guilmeau et al. 2017), with texturation leading to anisotropic properties with $\rho_{out\text{-}of\text{-}plane} > \rho_{in\text{-}plane}$ (1) and $\kappa_{out\text{-}of\text{-}plane} < \kappa_{in\text{-}plane}$ (2) (Figure 4.1), at all x, in agreement with measurements on crystals (Imai et al. 2001). Nevertheless, the most surprising result is the existence of an S anisotropy, with $|S_{in\text{-}plane}| > |S_{out\text{-}of\text{-}plane}|$ (3), a result differing from that reported for Bi_2Te_3 crystals that showed an isotropic S (Fleurial et al. 1988).

In Cu_xTiS_2, by combining the relations (1) and (3), the in-plane PF is always higher than the out-of-plane one, but when calculating zT, it appears that this in-plane superiority is always compensated by the relation (2), leading to similar zT values for in-plane and out-of-plane orientations (Guilmeau et al. 2017).

In other words, for TiS_2 and its related compounds, there is no advantage in texturing the material, as the zT values along the in-plane or out-of-plane directions are always similar. This explains why the zT values of crystals and polycrystals of that material are the same.

To conclude that chapter on TiS_2, it must be added that other strategies have been developed to take advantage of the 2D feature of its structure with a VDW gap. Misfit compounds of the $(MS)_{1+x}(TiS_2)_2$ formula (with M = Pb, Sn, Bi) have also been investigated (Wan et al. 2010, Guilmeau et al. 2015). The same kind of conclusions were drawn, the intergrown $(MS)_{1+x}$ extra layers being responsible for a decrease of the thermal conductivity, but by charge compensation between $(MS)_{1+x}$ and $(TiS_2)_2$ layers, a PF reduction was also measured. Successful grafting of organic chains in between the $[TiS_2]_\infty$ successive layers to produce hybrid TE materials was also shown to be promising: for instance, the hybrid inorganic–organic superlattice of TiS_2 $[\text{tetrabutylammonium}]_x$ was proposed as a flexible thermoelectric (Wan et al. 2017).

4.3. Magnetism and thermoelectricity in sulfides

The Seebeck coefficient can be written as the sum of two terms, the transport term (depending on the energy and charge currents) and the entropic term associated with the chemical potential (Peterson et al. 2010). The magnetic field dependence of thermopower has been actively investigated in the field of spintronics, and large magnetothermopower (MTEP) can be observed simultaneously with large magnetoresistance (MR) in Fe/Cr or Co/Cu multilayers (Sakurai et al. 1991; Piraux et al. 1992). For these metallic and spin-polarized multilayers, the MTEP values depend on the relative ratio between spin-up and spin-down channels (Piraux et al. 1992) following Mott's formula. Large enhancements of zT induced by a magnetic field have also been achieved in metallic nanowires such as Ni (with $\Delta zT/zT = +22\%$ in 0.5 T and 300 K), or with flexible Co/Cu nanowire networks (Mitdank et al. 2012; Araujo et al. 2019). More recently, large magnetothermopower values have been evidenced in topological insulators (Qu et al. 2012) induced by the peculiar band structures of these materials and associated with extremely large magneto-resistance values. All these effects are mostly due to the transport term of the Seebeck coefficient.

In the case of oxides such as Na_xCoO_2 (Wang et al. 2003) and misfits cobaltites (Maignan et al. 2003), or sulfides and selenides (Peix et al. 1985; Berthebaud et al. 2018), large magnetothermopower values have been obtained, but mostly due to the entropic part of thermopower. The presence of paramagnetic spins leads to a large magnetic entropy term, which can be quenched by the magnetic field application. This term can reach values as large as 60 μV/K in BiCaCoO misfits, with a relative magnetothermopower reaching -70% at 2 K and 9 T (Maignan et al. 2003). This contribution can be observed in a large T range, i.e. below 75 K in $CuCrTiS_4$ (Berthebaud et al. 2018) or below 100 K in misfits (Maignan et al. 2003).

Apart from this magnetic entropy term arising from the paramagnetic spins, a second entropic term has been evidenced in the case of narrow band systems such as Na_xCoO_2 and misfits, following the generalized Heikes formula. At sufficiently high temperature (i.e. T >> characteristic energy scales such as t, the hopping parameter), the atomic limit can be reached and the Seebeck coefficient is equal to the chemical potential, and thus depends on the spin and orbital entropy of the different transition metal cations inside the material (Doumerc 1994; Koshibae et al. 2001). Large values for this entropic term have been reported and considered to explain the large thermoelectric power, as in Na_xCoO_2 (Koshibae et al. 2001; Wang et al. 2003).

Magnetism is therefore an interesting tuning parameter of the transport and thermoelectric properties. The entropic term can be more easily observed in oxides, due to the narrow electronic band induced by electronic correlations (Koshibae et al. 2001; Peterson et al. 2007), but sulfides can also exhibit a large magnetothermopower or a large interplay between thermoelectric coefficients and magnetism. We review here recent results obtained in sulfides to show the positive impact of magnetism on the Seebeck coefficient in the presence of spin-polarized transport or with paramagnetic spins.

The thermopowers of two binary cobalt sulfides are presented in Figure 4.2.

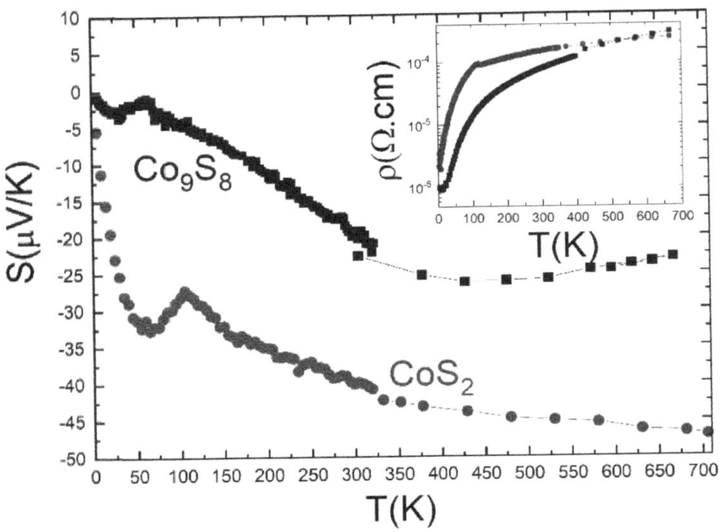

Figure 4.2. *The Seebeck coefficient of CoS_2 and Co_9S_8 as a function of temperature. Inset: electrical resistivity of CoS_2 and Co_9S_8*

The Seebeck coefficients of Co_9S_8 and CoS_2 are negative with values reaching –25 μV/K and -50 μV/K, respectively, at 700 K. Co_9S_8 is metallic with very small values of electrical resistivity (inset of Figure 4.2), close to 0.08 mΩ.cm at 300 K, and is reported to be antiferromagnetic with a very large T_N (Heidelberg et al. 1966), or a Pauli paramagnet (Covino et al. 1982), i.e. no magnetic transition is observed for the two cases up to at least 1,000 K. Only few reports can be found on this material, and its magnetic properties are still debated. The S values are almost linear up to 300 K, with relatively modest values in good agreement with its metallic behavior. The electrical resistivity follows a Bloch Gruneisen behavior suggesting electron–phonon-dominated transport (Hébert et al. 2021b). Contrary to Co_9S_8, CoS_2 has been deeply investigated and can be considered as a model system for spin-polarized transport, with its ferromagnetic behavior below T_C = 122 K coexisting with metallicity. Spin polarization is very large and can reach 85% by the Fe for Co substitution (Leighton et al. 2007). The strong coupling between charges and spins leads to a clear accident on the ρ(T) curve at 122 K (inset of Figure 4.2). The S values are much larger for CoS_2 throughout the T range, and moreover a peak of S (Figure 4.2) is observed in the ferromagnetic state below 122 K, which can be fitted by a

$T^{3/2}$ behavior (Figure 4.3) reminiscent of a magnonic contribution (Hébert et al. 2013b).

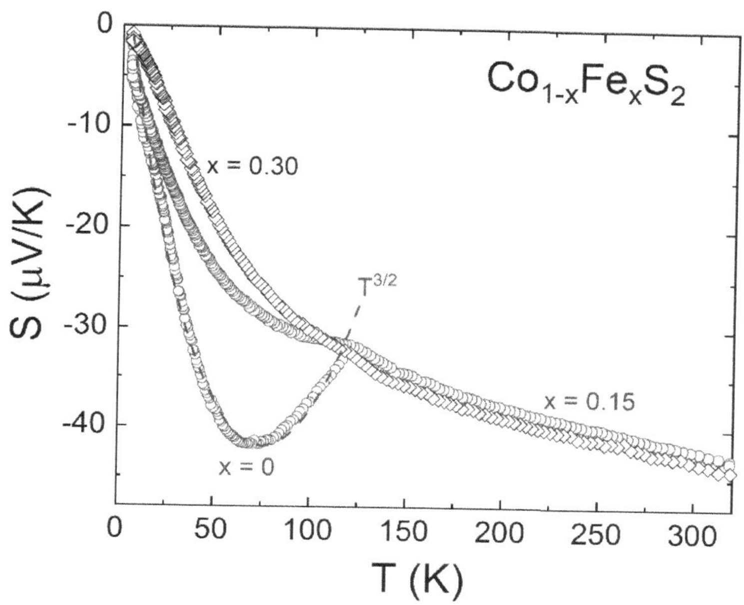

Figure 4.3. *The Seebeck coefficient of $Co_{1-x}Fe_xS_2$. The dashed line is a fit of $S \sim a+bT+cT^{3/2}$ extracted from Hébert (2013b)*

The absolute values of S steadily increase up to 700 K, and almost saturate to S = -50 μV/K. The resultant power factor S^2/ρ is very large and moreover strongly increased when entering into the FM state, exceeding 5 $mWm^{-1}K^{-2}$ at low temperatures (Hébert et al. 2013b). The impact of Fe substitution can be observed in Figure 4.3.

As Fe content increases, the low T peak of thermopower gradually decreases, while the values above T_C remain unaffected. The peak is completely suppressed for x > 0.15 in $Co_{1-x}Fe_xS_2$ (Acevedo Salas et al. 2019). This critical x value directly corresponds to the value where spin polarization reaches its maximal value (Leighton et al. 2007), suggesting that the increase of the imbalance between spin-up and spin-down channels as x decreases from x = 0.15 is at the origin of the peak. The tuning of spin polarization in such a metallic material can thus generate very large Seebeck values and very large power factors.

Magnetism can also play a role in the thermoelectric properties in the case of localized carriers. In the thiospinel $CuCrTiS_4$, a spin glass behavior has been reported at T = 8 K (Nagata et al. 2012). A deviation from the Curie–Weiss law is observed below T ~ 75 K (Figure 4.5 of Berthebaud et al. (2018)). Resistivity values are typical of localized carriers (Figure 4.4), and the ρ(T) curve displays a variable-range-hopping behavior below 75 K. Remarkably, very large values of magneto-resistance (MR) are observed below ~ 100 K, with MR reaching -50% at 20 K and 9 T (Figure 4.5, left axis) and up to -95% at 5 K/9 T (Berthebaud et al. 2018). All the MR curves can be rescaled on a single curve following the Brillouin function as $(1-B_J^2)$, where B_J is the Brillouin function with J = 3/2. This scaling is consistent with the hopping of Cr^{3+} (S = 3/2), and with the strong reduction of ρ as Cr^{3+} spins align under the application of a magnetic field. This is reminiscent of the behavior of colossal magneto-resistant manganites (Viret et al. 1997).

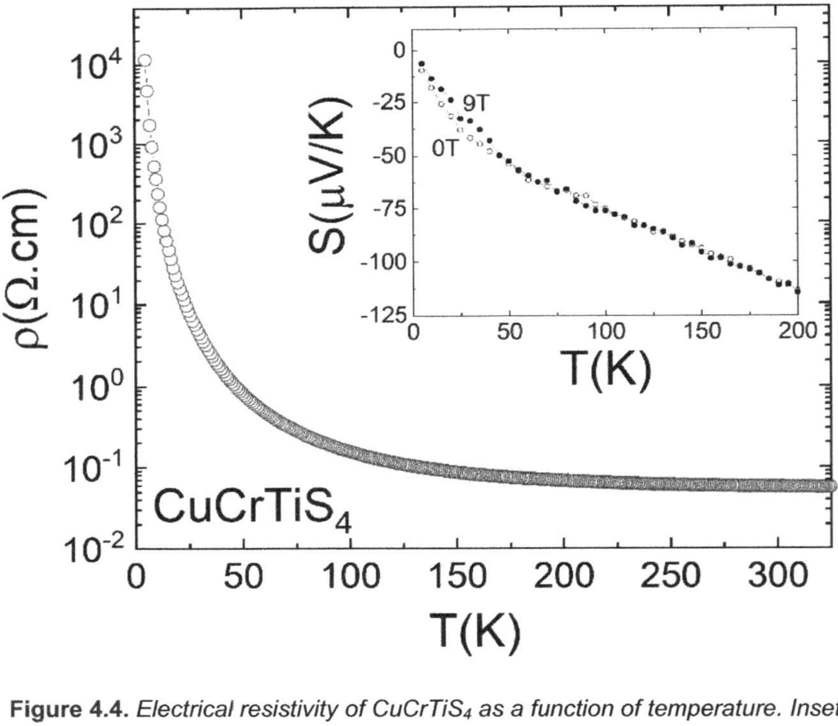

Figure 4.4. *Electrical resistivity of $CuCrTiS_4$ as a function of temperature. Inset: Seebeck coefficient as a function of temperature, measured under 0 T and 9 T*

Such a large MR has previously been observed in $FeCr_2S_4$ thiospinel (Ramirez et al. 1997), and contrasts with the behavior of a metallic and Pauli paramagnetic thiospinel such as $CuTi_2S_4$, in which a small and positive MR is observed (Berthebaud et al. 2018). The inset of Figure 4.4 presents the thermopower of $CuCrTiS_4$ measured in 0 T and 9 T. The values are negative and continuously increase in magnitude as T increases. The two curves are superimposed at high temperatures, but below ~75 K, the two curves start to separate, with a larger magnitude in 0 T than in 9 T.

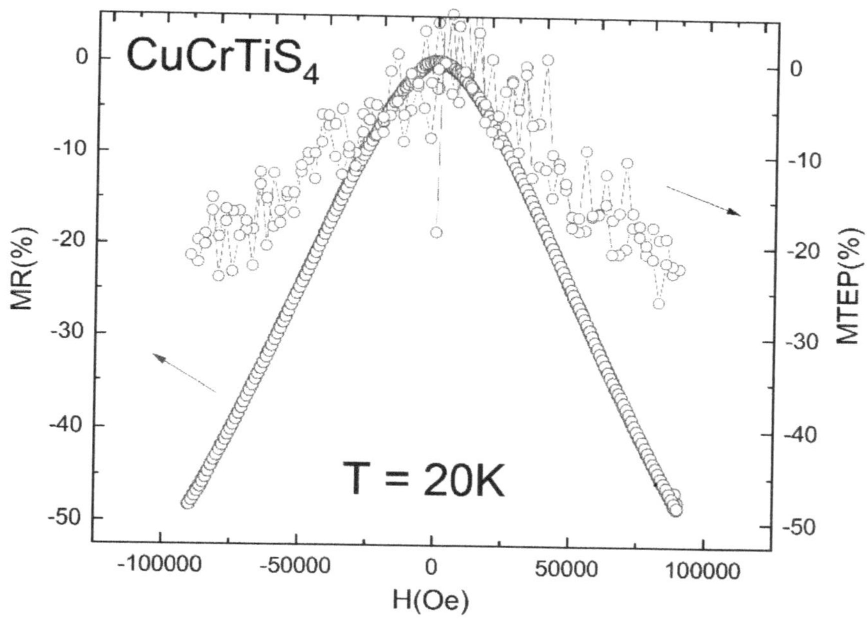

Figure 4.5. *Magneto-resistance MR (left-axis) and magneto-thermopower MTEP (right-axis) as a function of magnetic field at T = 20 K in $CuCrTiS_4$*

As shown in Figure 4.5 (right-axis), the magnitude of S at 20 K continuously decreases as a function of field, with magnetothermopower values reaching –26% at 5 K and 9 T. For these low temperatures, the very large MR (Figure 4.5, left-axis) of -50% induces a positive impact on the power factor PF, even if MTEP is negative, as PF increases by 15%. This has no direct impact on applications as the temperatures and PF values are much too small, but the combined mechanism of large and negative MR together with a negative MTEP, smaller in magnitude, can be interesting for materials exhibiting MR values closer to room temperature. The link between MR and MTEP is still under investigation.

4.4. Conclusion

Two different strategies have been presented here to show how the thermoelectric properties of sulfides can be modified or optimized. A careful doping investigation has been performed in TiS_2 single crystals. Even if the thermal and electronic transport are very anisotropic, the zT values do not so strongly depend on the anisotropy, showing that texturation is not necessary in this family of materials. In magnetic sulfides, the impact of magnetic fields and magnetism can be important both for itinerant and localized carriers, and magnetism can induce an enhancement of the Seebeck coefficient. Magnetism can be considered as an extra parameter, with respect to doping, nanostructuration or microstructure optimization, for the optimization of thermoelectric properties in these sulfides.

4.5. References

Acevedo Salas, U., Fourati, I., Juraszek, J., Richomme, F., Pelloquin, D., Maignan, A., Hébert, S. (2019). Impact of the iron substitution on the thermoelectric properties of $Co_{1-x}Fe_xS_2$ ($x \leq 0.30$). *Phil. Trans. R. Soc. A*, 377, 20180337. doi: 10.1098/rsta.2018.0337.

Araujo, F.A., da Camara Santa Clara Gomes, T., Piraux, L. (2019). Magnetic control of flexible thermoelectric devices based on macroscopic 3D interconnected nanowire networks. *Adv. Elec. Mater.*, 5, 1800819.

Barbier, T., Lebedev, O.I., Roddatis, V., Bréard, Y., Maignan, A., Guilmeau, E. (2015). Silver intercalation in SPS dense TiS_2: Staging and thermoelectric properties. *Dalton Trans.*, 44, 7887. doi: 10.1039/C5DT00551E.

Beaumale, M., Barbier, T., Bréard, Y., Guelou, G., Powell, A.V., Vaqueiro, P., Guilmeau, E. (2014). Electron doping and phonon scattering in $Ti_{1+x}S_2$ thermoelectric compounds. *Acta Mater.*, 78, 86. doi: 10.1016/j.actamat.2014.06.032.

Behnia, K., Jaccard, D., Flouquet, J. (2004). On the thermoelectricity of correlated electrons in the zero-temperature limit. *J. Phys. Condens. Matter*, 16, 5187–5198. doi: 10.1088/0953-8984/16/28/037.

Berthebaud, D., Lebedev, O.I., Maignan, A., Hébert, S. (2018). Magnetothermopower and giant magnetoresistance in the spin-glass $CuCrTiS_4$ thiospinel. *J. Appl. Phys.*, 124, 063905. doi: 10.1063/1.5036828.

Covino, J., Pasquariello, D., Kim, K., Dwight, K., Wold, A. (1982). Magnetic properties of bulk and supported Co_9S_8. *Mater. Res. Bull.*, 17, 1191 and references therein.

Daou, R., Takahashi, H., Hébert, S., Guilmeau, E., Maignan, A. (2015). Intrinsic effects of substitution and intercalation on thermal transport in two-dimensional TiS_2 single crystals. *J. Appl. Phys.*, 117, 165101. doi: 10.1063/1.4919078.

Doumerc, J.-P. (1994). Thermoelectric power for carriers in localized states: A generalization of heikes and chaikin-beni formulae. *J. Solid Stat. Chem.*, 110, 419.

Fleurial, J.-P., Gailliard, L., Triboulet, R., Scherrer, H., Scherrer, S. (1988). Thermal properties of high quality single crystals of bismuth telluride – Part I: Experimental characterization. *J. Phys. Chem. Solids*, 49, 1237. doi: 10.1016/0022-3697(88)90182-5.

Guélou, G., Vaqueiro, P., Prado-Gonjal, J., Barbier, T., Hébert, S., Guilmeau, E., Kockelmann, W., Powell A.V. (2016). The impact of charge transfer and structural disorder on the thermoelectric properties of cobalt intercalated TiS_2. *J. Mater. Chem. C*, 4, 1871. doi: 10.1039/c5tc04217h.

Guilmeau, E., Maignan, A., Martin, C. (2009). Thermoelectric oxides: Effect of doping in delafossites and zinc oxide. *J. Elec. Mater.*, 38, 1104–1108. doi: 10.1007/s11664-009-0815-2.

Guilmeau, E., Bréard, Y., Maignan, A. (2011). Transport and thermoelectric properties in copper intercalated TiS_2 chalcogenide. *Appl. Phys. Lett.*, 99, 052107. doi: 10.1063/1.3621834.

Guilmeau, E., Maignan, A., Wan, C., Koumoto, K. (2015). On the effects of substitution, intercalation, non-stoichiometry and block layer concept in TiS_2 based thermoelectrics. *Phys. Chem. Chem. Phys.*, 17, 24541–24555 [Online]. Available at: https://pubs.rsc.org/en/content/articlehtml/2015/cp/c5cp01795e.

Guilmeau, E., Barbier, T., Maignan, A., Chateigner, D. (2017). Thermoelectric anisotropy and texture of intercalated TiS_2. *Appl. Phys. Lett.*, 111, 133903. doi: 10.1063/1.4998952.

Hébert, S., Wang, L.B., Maignan, A., Pelloquin, D., Hervieu, M., Raveau, B. (2001). Large thermopower in metallic misfit cobalt oxides: Improvement by cationic substitutions. *MRS Online Proc. Library*, 691, 125. doi: 10.1557/PROC-691-G12.5.

Hébert, S., Kobayashi, W., Muguerra, H., Bréard, Y., Nunna, R., Gascoin, F., Guilmeau, E., Maignan, A. (2013a). From oxides to selenides and sulfides: The richness of the CdI_2 type crystallographic structure for thermoelectric properties. *Phys. Status Sol. A – Appl. and Mater. Sci.*, 210, 69. doi: 10.1002/pssa.201228505.

Hébert, S., Guilmeau, E., Bertebaud, D., Lebedev, O.I., Roddatis, V., Maignan, A. (2013b). Transport and magnetic properties of highly densified CoS_2 ceramic. *J. Appl. Phys.*, 114, 103703. doi: 10.1063/1.4820564.

Hébert, S., Daou, R., Maignan, A., Das, S., Banerjee, A., Klein, Y., Bourgès, C., Tsujii, N., Mori, T. (2021a). Thermoelectric materials taking advantage of spin entropy: Lessons from chalcogenides and oxides. *Sci. Technol. Adv. Mater.*, 22, 583–596. doi: 10.1080/14686996.2021.1951593.

Hébert, S., Barbier, T., Berthebaud, D., Lebedev, O.I., Pralong, V., Maignan, A. (2021b). Transport and thermoelectric coefficients of the Co_9S_8 metal: A comparison with the spin polarized CoS_2. *J. Phys. Chem. C*, 125, 5386–5391. doi: 10.1021/acs.jpcc.0c11601.

Heidelberg, R.F., Luxem, A.H., Talhouk, S., Banewicz, J.J. (1966). The magnetic susceptibilities of the cobalt-sulfur system. *Inorg. Chem.*, 5, 194.

Imai, H., Shimakawa, Y., Kubo, Y. (2001). Large thermoelectric power factor in TiS_2 crystal with nearly stoichiometric composition. *Phys. Rev. B*, 64, 241104. doi: 10.1103/PhysRevB.64.241104.

Kawano, K., Takahashi, J., Nakata, R., Sumita, M. (1989). Temperature dependence of resistivities in TiS_2 and its Cu-intercalate. *J. Phys. Soc. Jpn*, 58, 949. doi: 10.1143/JPSJ.58.949.

Koshibae, W. and Maekawa, S. (2001). Effects of spin and orbital degeneracy on the thermopower of strongly correlated systems. *Phys. Rev. Lett.*, 87, 236603. doi: 10.1103/PhysRevLett.87.236603.

Koumoto, K., Wang, Y., Zhang, R., Kosuga, A., Funahashi, R. (2010). Oxide thermoelectric materials: A nanostructuring approach, annual review of materials research. *Annu. Rev. Mater. Res.*, 40, 363–394. doi: 10.1146/annurev-matsci-070909-104521.

Leighton, C., Manno, M., Cady, A., Freeland, J.W., Wang, L., Umemoto, K., Wentzcovitch, R.M., Chen, T.Y., Chien, C.L., Kuhns, P.L. et al. (2007). Composition controlled spin polarization in $Co_{1-x}Fe_xS_2$ alloys. *J. Phys. Condens. Matter*, 19, 315219. doi: 10.1088/0953-8984/19/31/315219.

Li, J., Tan, Q., Li, J.F. (2013). Synthesis and property evaluation of $CuFeS_{2-x}$ as earth-abundant and environmentally-friendly thermoelectric materials. *J. Alloys Compd.*, 551, 143–149. doi: 10.1016/j.jallcom.2012.09.067.

Maignan, A., Hébert, S., Hervieu, M., Michel, C., Pelloquin, D., Khomskii, D. (2003). Magnetoresistance and magnetothermopower properties of Bi/Ca/Co/O and Bi(Pb)/Ca/Co/O misfit layer cobaltites. *J. Phys. Condens. Matter*, 15, 2711–2723. doi: 10.1088/0953-8984/15/17/323.

Mangelis, P., Vaqueiro, P., Powell, A.V. (2020). Improved thermoelectric performance through double substitution in shandite-type mixed-metal sulfides. *ACS Appl. Energy Mater.*, 3, 2168–2174. doi: 10.1021/acsaem.9b02272.

Masset, A.C., Michel, C., Maignan, A., Hervieu, M., Toulemonde, O., Studer, F., Raveau, B., Hejtmanek, J. (2000). Misfit-layered cobaltite with an anisotropic giant magnetoresistance: $Ca_3Co_4O_9$. *Phys. Rev. B*, 62, 166–175. doi: 10.1103/PhysRevB.62.166.

Mitdank, R., Handwerg, M., Steinweg, C., Töllner, W., Daub, M., Nielsch, K., Fischer, S.F. (2012). Enhanced magneto-thermoelectric power factor of a 70nm Ni-nanowire. *J. Appl. Phys.*, 111, 104320.

Nagata, S., Koseki, N., Ebisu, S. (2012). Spin-glass in the spinel-type $CuCrTiS_4$. *Philos. Mag. B*, 92, 2957.

Ong, K.P., Singh, D.J., Wu, P. (2010). Unusual transport and strongly anisotropic thermopower in $PtCoO_2$ and $PdCoO_2$. *Phys. Rev. Lett.*, 104, 176601. doi: 10.1103/PhysRevLett.104.176601

Pawula, F., Daou, R., Hébert, S., Lebedev, O., Maignan, A., Subedi, A., Kakefuda, Y., Kawamoto, N., Baba, T., Mori, T. (2019). Anisotropic thermal transport in magnetic intercalates Fe_xTiS_2. *Phys. Rev. B*, 99, 085422. doi: 10.1103/PhysRevB.99.085422.

Peix, G., Babot, G., Chevreton, M. (1985). Etude magnétothermoélectrique de la transition métal-semiconducteur dans les composés $Cr_{2+\varepsilon}Se_3$ ($0 \leq x \leq 0.44$). *J. Solid Stat. Chem.*, 56, 304.

Peterson, M.R. and Shastry, S. (2010). Kelvin formula for thermopower. *Phys. Rev. B*, 82, 195105.

Peterson, M.R., Mukerjee, S., Sriram Shastry, B., Haerter, J.O. (2007). Dynamical thermal response functions for strongly correlated one-dimensional systems: Hubbard and spinless fermion t–V model. *Phys. Rev. B*, 76, 125110. doi: 10.1103/PhysRevB.76.125110.

Piraux, L., Fert, A., Schroeder, P.A., Loloee, R., Etienne, P. (1992). Large magnetothermoelectric power in Co/Cu, Fe/Cu and Fe/Cr multilayers. *J. Mag. Mag. Mater.*, 110, L274.

Powell, A.V. (2019). Recent developments in Earth-abundant copper-sulfide thermoelectric materials, *J. Appl. Phys.*, 126, 100901. doi: 10.1063/1.5119345.

Qu, D.-X., Hor, Y.S., Cava, R.J. (2012). Quantum oscillations in magnetothermopower measurements of the topological insulator Bi_2Te_3. *Phys. Rev. Lett.*, 109, 246502.

Ramirez, A.P., Cava, R., Krajewski, J. (1997). Colossal magnetoresistance in Cr-based chalcogenide spinels. *Nature*, 386, 156.

Sakurai, J., Horie, M., Haraki, S., Yamamoto, M., Shinjo, T. (1991). Magnetic fields effects on thermopower of Fe/Cr, Cu/Co and Cu/Ni(Fe) multilayers. *J. Phys. Soc. Jpn*, 60, 2522.

Scherrer, H. and Scherrer, S. (1995). Bismuth telluride, antimony telluride, and their solid solutions. In *CRC Handbook of Thermoelectrics*, Rowe, D.M. (ed.). CRC Press, Boca Raton.

Slack, G. (1997). Design concepts for improved thermoelectric materials. *MRS Proc.*, 478, 47. doi: 10.1557/PROC-478-47.

Sotelo, A., Rasekha, S.H., Guilmeau, E., Madrea, M.A., Torres, M.A., Marinel, S., Diez, J.C. (2011). Improved thermoelectric properties in directionally grown $Bi_2Sr_2Co_{1.8}O_y$ ceramics by Pb for Bi substitution. *Mater. Res. Bull.*, 46, 2537. doi: 10.1016/j.materresbull.2011.08.011.

Stoliaroff, A., Jobic, S., Latouche, C. (2019). Optoelectronic properties of TiS_2: A never ended story tackled by density functional theory and many-body methods. *Inorg. Chem.*, 58, 1949–1957. doi: 10.1021/acs.inorgchem.8b02883.

Torres, M.A., Costa, F.M., Flahaut, D., Touati, K., Rasekh, S.H., Ferreira, N.M., Allouche, J., Depriester, M., Madre, M.A., Kovalevsky, A.V. et al. (2019). Significant enhancement of the thermoelectric performance in $Ca_3Co_4O_9$ thermoelectric materials through combined strontium substitution and hot-pressing process. *J. Eur. Ceram. Soc.*, 39, 1186. doi: 10.1016/j.jeurceramsoc.2018.12.049.

Viret, M., Ranno, L., Coey, J.M.D. (1997), Magnetic localization in mixed-valence manganites. *Phys. Rev. B*, 55, 8067.

Wan, C., Wang, Y., Wang, N., Koumoto, K. (2010). Low-thermal-conductivity $(MS)_{1+x}(TiS_2)_2$ (M = Pb, Bi, Sn) misfit layer compounds for bulk thermoelectric materials. *Materials*, 3, 2606. doi: 10.3390/ma3042606.

Wan, C., Tian, R., Kondou, R.Y.M., Zong, P., Koumoto, K. (2017). Ultrahigh thermoelectric power factor in flexible hybrid inorganic-organic superlattice. *Nature Communications*, 8, 1024. doi: 10.1038/s41467-017-01149-4.

Wang, Y., Rogado, N.S., Cava, R.J., Ong, N.P. (2003). Spin entropy as the likely source of enhanced thermopower in $Na_xCo_2O_4$. *Nature*, 423, 425–428. doi: 10.1038/nature01639.

5

Thermoelectric Generation Using the Anomalous Nernst Effect

Akito SAKAI[1] and Satoru NAKATSUJI[1–5]

[1] Department of Physics, University of Tokyo, Japan
[2] Institute for Solid State Physics, University of Tokyo, Japan
[3] CREST, Japan Science and Technology Agency (JST), Saitama, Japan
[4] Trans-scale Quantum Science Institute, University of Tokyo, Japan
[5] Institute for Quantum Matter and Department of Physics and Astronomy, Johns Hopkins University, Baltimore, USA

5.1. Thermoelectric conversion – Seebeck effect and anomalous Nernst effect (ANE)

Thermoelectric (TE) conversion, the direct generation of electricity from heat current, has been extensively studied for more than a century for using waste heat in our environment as a source of energy. In addition, today in powering trillions of IoT sensors and wearable devices, there is a rapidly growing demand for novel energy harvesting technology instead of batteries, in particular in the form of flexible, durable thermoelectric generators. Although the thermoelectric conversion has been regarded as a leading technology, the practical application using the thermopile based on the

For a color version of all the figures in this chapter, see www.iste.co.uk/akinaga/thermoelectric2.zip.

Thermoelectric Micro/Nano Generators 2,
coordinated by Hiroyuki AKINAGA, Atsuko KOSUGA,
Takao MORI and Gustavo ARDILA.

conventional Seebeck effect faces a number of difficulties such as very high production cost (for the complicated pillar structure and a large number of contacts (Figure 5.1a)) and the usage of the rare, toxic and brittle elements such as Bi, Pb and Te. Thus, another type of thermoelectric conversion, the anomalous Nernst effect (ANE), has attracted significant attention as the next-generation technology (Mizuguchi and Nakatsuji 2019): its transverse geometry allows the efficient, large area and flexible coverage of a curved heat source with a much lower production cost (Figure 5.1b). Moreover, the recent discovery of the significant enhancement in the ANE in topological magnets has initiated a new trend in the study of quantum materials and paved a path for its TE application (Ikhlas and Tomita 2017; Sakai et al. 2018, 2020; Nakatsuji and Arita 2022). Thus, the large ANE in topological magnets has gained much interest not only from academia but also from industrial sectors because of the potential to reduce the production cost of thermopiles by two to three orders of magnitude, enabling the design of low-cost energy-harvesters for IoT sensors and wearable devices.

Firstly, let us review the Seebeck and anomalous Nernst effects. Up until now, most of the TE conversion is based on the Seebeck effect, which is a longitudinal TE response (Figure 5.1a) (Nolas et al. 2001).

$$S_{xx} \equiv \frac{E_x}{(-\partial T/\partial x)} = \frac{\Delta V}{\Delta T}. \quad [5.1]$$

Here, S_{xx}, $\boldsymbol{E}$, $\boldsymbol{\nabla}T$, ΔV, and ΔT are the Seebeck coefficient, electric field, temperature gradient, electromotive force and temperature difference, respectively. Note that ΔV does not depend on the material's geometry ℓ, since $\boldsymbol{E}$ and $\boldsymbol{\nabla}T$ are parallel to each other. This effect indicates that the carriers in the material (electron or hole) can carry both the charge and heat current, which is generalized by the Boltzmann transport equations,

$$\boldsymbol{J}_c = \hat{\sigma}\boldsymbol{E} + \hat{\alpha}(-\boldsymbol{\nabla}T) \quad [5.2]$$

$$\boldsymbol{J}_q = \hat{\Pi}\boldsymbol{J}_c + \hat{\kappa}(-\boldsymbol{\nabla}T). \quad [5.3]$$

Here, $\boldsymbol{J}_c$, $\boldsymbol{J}_q$, $\hat{\sigma}$, $\hat{\alpha}$, $\hat{\Pi}$, and $\hat{\kappa}$ are the electric current density, heat current density, electrical conductivity tensor, thermoelectric conductivity tensor, the Peltier tensor and thermal conductivity tensor, respectively. The first term in equation [5.2] and the second term in equation [5.3], respectively, represent Ohm's law and Fourier's law. TE conversion is described by the second term in equation [5.2] and the first term in equation [5.3]. Assuming time-reversal symmetry ($\sigma_{ij} = \alpha_{ij} = 0\ (i \neq j)$) and setting $\boldsymbol{J}_c = 0$ and $-\boldsymbol{\nabla}T \parallel \boldsymbol{x}$, we can obtain $S_{xx} = E_x/(-\partial T/\partial x) = \alpha_{xx}/\sigma_{xx}$.

If the external magnetic field is applied or the material itself shows a ferromagnetic order, the off-diagonal elements can be nonzero. Especially for the ferromagnetic case, α_{ij} and σ_{ij} can be finite even at zero external magnetic fields owing to the spontaneous magnetization. Assuming the z-axis ferromagnetic moment $\boldsymbol{M} \parallel \boldsymbol{z}$ and setting $\boldsymbol{J}_c = 0$ and $-\boldsymbol{\nabla}T \parallel \boldsymbol{x}$, we can obtain:

$$J_y = \sigma_{yx}S_{xx} + \sigma_{yy}S_{yx} - \alpha_{yx} = 0, \qquad [5.4]$$

where S_{yx} is the anomalous Nernst (AN) coefficient, which is defined by,

$$S_{yx} \equiv \frac{E_y}{(-\partial T/\partial x)} = \frac{t}{\ell}\frac{\Delta V}{\Delta T}. \qquad [5.5]$$

Here, ℓ and t are the sample lengths along the direction of temperature gradient and along the electromotive force, respectively (Figure 5.1b). In contrast with S_{xx}, S_{yx} depends on the sample geometry ℓ and t and more precisely scales with t/ℓ because $\boldsymbol{E} \perp \boldsymbol{\nabla}T$. As a result, the device structure for the ANE has to be different from that for the Seebeck effect. Namely, the conventional thermoelectric device (Peltier device) consists of a large number of p- and n-type semiconductor pillars to enhance the voltage (Figure 5.1a), while the AN device may have a flexible thin-film structure, which can efficiently cover curved and wide heat sources (Figure 5.1b) (Sakuraba et al. 2013). However, the $|S_{yx}|$ for the ordinary ferromagnet is ~ 0.1% of $|S_{xx}|$, and is too small for any thermoelectric applications. In the following section, we describe the relation between S_{yx} and the electronic band structure, and the recent observation of the giant ANE at room temperature found in topological magnets (Nakatsuji and Arita 2022).

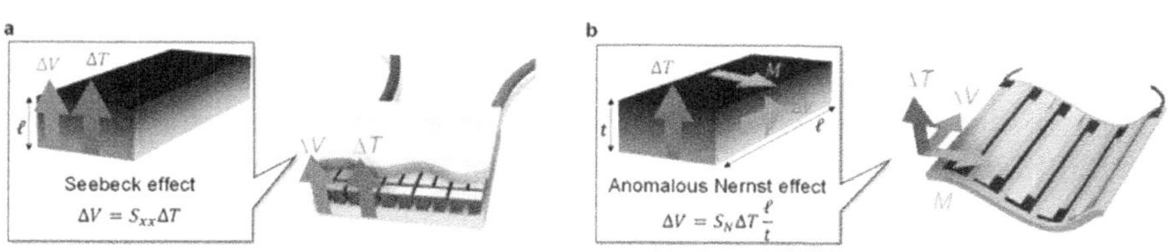

Figure 5.1. *Seebeck effect and anomalous Nernst effect*

5.2. Physics of topological magnets

The electronic band structure, in particular, the bands near Fermi energy, ε_F governs the electronic transport properties of a conductor. Topological magnets possess both magnetic ordering and unique topological band structure near ε_F such as linear crossings (Weyl cones) and touching in lines (nodal lines). These touching points or lines in the band structure enhance the Berry curvature, which can induce a large anomalous Hall and anomalous Nernst effect (Nakatsuji and Arita 2022).

5.2.1. *Transverse electrical and thermal conductivity driven by Berry curvature*

The transverse electrical and thermal conductivity, σ_{xy} and α_{xy}, are obtained using the Kubo formula,

$$\sigma_{xy}(T,\varepsilon_F) = -\frac{e^2}{\hbar}\int \frac{d^D k}{(2\pi)^D} \sum_n f(T,\varepsilon_{n,\boldsymbol{k}})\Omega_{n,z}(\boldsymbol{k}), \tag{5.6}$$

$$\alpha_{xy}(T,\varepsilon_F) = -\frac{1}{e}\int d\varepsilon\sigma_{xy}(0,\varepsilon)\frac{\varepsilon-\varepsilon_F}{T}\left(-\frac{\partial f}{\partial \varepsilon}\right), \tag{5.7}$$

$$= \frac{k_B}{e}\int d\varepsilon \left(\frac{\partial \sigma_{xy}}{\partial \varepsilon}\right) s(\varepsilon,T), \tag{5.8}$$

where D is the dimension of the system, f is the Fermi–Dirac distribution function, and $s = -f\ln f - (1-f)\ln(1-f)$ is the entropy density. $\boldsymbol{\Omega}_n(\boldsymbol{k}) \equiv \boldsymbol{\nabla}_k \times \boldsymbol{A}_n(\boldsymbol{k})$ is the Berry curvature, where $\boldsymbol{A}_n(\boldsymbol{k}) \equiv -i\langle n(\boldsymbol{k})|\boldsymbol{\nabla}_k|n(\boldsymbol{k})\rangle$ is the Berry connection of the n-th band $|n(\boldsymbol{k})\rangle$. If $D = 2$ and the Fermi energy ε_F lies within the band gap such as two-dimensional (2D) electron systems under magnetic field and 2D band insulators, the integration of equation [5.6] is over the entire Brillouin zone, resulting in the integer Hall effect $\sigma_{xy} = -e^2 C_n/h$, if the Chern number C_n is nonzero and an integer value (Thouless et al. 1982). On the contrary, equation [5.6] is integrated over ε_F for 3D metals, resulting in unquantized σ_{xy} called intrinsic anomalous Hall conductivity. By using the partial derivative of the Hamiltonian $\boldsymbol{\nabla}_k\mathcal{H}$, the Berry curvature can also be expressed as,

$$\boldsymbol{\Omega}_n(\boldsymbol{k}) = i \sum_{m \neq n} \frac{\langle n(\boldsymbol{k})|\nabla_k \mathcal{H}|m(\boldsymbol{k})\rangle \times \langle m(\boldsymbol{k})|\nabla_k \mathcal{H}|n(\boldsymbol{k})\rangle}{(\varepsilon_n - \varepsilon_m)^2}. \qquad [5.9]$$

The magnitude of Berry curvature becomes gigantic around the touching points or lines since the denominator of equation [5.9] is the energy difference of the two bands. Then, in turn, such large Berry curvature enhances σ_{xy} and α_{xy} through equations [5.6] and [5.7] (Nagaosa et al. 2010; Xiao et al. 2010). We note that to obtain large α_{xy} since $\frac{\partial \sigma_{xy}}{\partial \varepsilon} = \epsilon_{ijl} \sum_{n,\boldsymbol{k}} \Omega_{n,l}(\boldsymbol{k}) \delta(\varepsilon - \varepsilon_{n,\boldsymbol{k}})$, it is important to simultaneously enhance both the Berry curvature and the density of state (DOS) of the band generating the Berry curvature. In fact, such a situation is realized in Co_2MnGa and Fe_3X (X = Ga, Al) as we shall describe in section 5.3.

5.2.2. *Magnetic Weyl semimetals, Weyl magnets*

Weyl semimetals possess Weyl points near ε_F, which are the touching points of two bands with a linear dispersion and described by the following Weyl Hamiltonian,

$$\mathcal{H}(\boldsymbol{k}) = \gamma(\boldsymbol{k} - \gamma\, \boldsymbol{k_0}) \cdot \boldsymbol{\sigma} = \gamma \sum_{i=x}^{z} (k_i - \gamma k_{0,i}) \sigma_i, \qquad [5.10]$$

where $\gamma = \pm 1$, $\boldsymbol{k}$, $\pm \boldsymbol{k_0}$, and σ_i, respectively, denote a monopole charge (or helicity, chirality), the wave vector, the position of the Weyl points and i-th Pauli matrix. We can easily obtain its eigen energies,

$$\varepsilon_\pm = \pm|\boldsymbol{k} - \gamma\, \boldsymbol{k_0}| \equiv \pm|\delta \boldsymbol{k}|, \qquad [5.11]$$

which express cone-shaped dispersion touching at the doubly degenerate Weyl points $\boldsymbol{k} = \pm \boldsymbol{k_0}$ (Figure 5.2a). The Weyl points are sources ($\gamma = 1$) or sinks ($\gamma = -1$) of Berry curvature and are always created/annihilated in pairs of opposite chiralities (Nielsen–Ninomiya theorem) (Nielsen and Ninomiya 1981). If the system has both time-reversal symmetry (TRS) and space inversion symmetry (IS), the Nielsen–Ninomiya theorem assures the two Weyl points with opposite chiralities at the same $\boldsymbol{k_0}$ position, resulting in a four-fold degenerate Dirac point. Thus, either TRS or IS is broken in Weyl semimetals. Especially, Weyl magnets (or magnetic Weyl semimetals) break TRS by a magnetic order (Wan et al. 2011; Armitage et al. 2018).

5.2.3. *Type-II Weyl semimetals*

The Weyl equation, first appeared in quantum field theory, is originally invariant under the Lorentz transformation. In condensed matter physics, however, the Weyl equation represents a model Hamiltonian of the electronic band structure, and the Lorentz invariance can be violated. Let us consider the following Hamiltonian as an example:

$$\mathcal{H}(\boldsymbol{k}) = \gamma\hbar v_2(k_z \mp k_0)\sigma_0 + \hbar v_\perp(k_x\sigma_x + k_y\sigma_y) + \gamma\hbar v_1(k_z \mp k_0)\sigma_z. \qquad [5.12]$$

The second and third terms represent Weyl point at $\pm\boldsymbol{k}_0 = (0,0,\pm k_0)$, and the first term gives the tilt of Weyl cone. The eigen energy of equation [5.12] is:

$$\varepsilon_\pm = \hbar v_2 \delta k_z \pm \sqrt{(\hbar v_\perp)^2 k_\perp^2 + (\hbar v_1)^2 \delta k_z^2}, \qquad [5.13]$$

where $k_\perp^2 \equiv k_x^2 + k_y^2, \delta k_z \equiv k_z - \gamma k_0$. This energy dispersion is illustrated in Figure 5.2b. For $|v_2/v_1| < 1$ and setting $\varepsilon_F = 0$, only the Weyl points are located at ε_F (type-I Weyl). On the contrary, the tilted Weyl cones cut the Fermi energy $\varepsilon_F = 0$ for $|v_2/v_1| > 1$, resulting in electron and hole pockets touching at the Weyl points (type-II Weyl) (Soluyanov et al. 2015). Thus, a topological phase transition of Fermi surface occurs at $|v_2/v_1| = 1$. Here, the topology of the Fermi surfaces chagnes across $|v_2/v_1| = 1$. This transition can be viewed as a quantum Lifshitz transition.

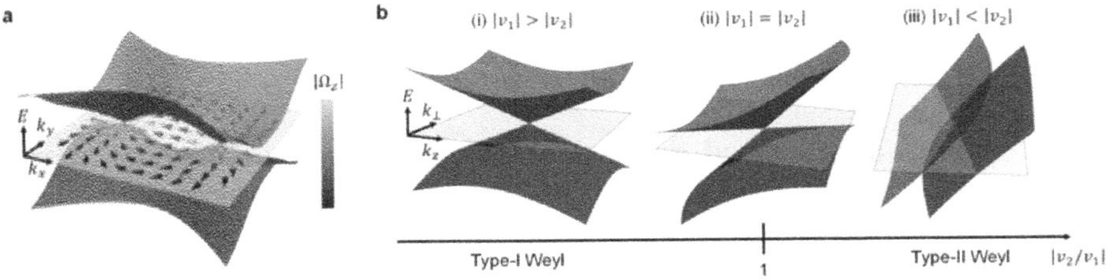

Figure 5.2. *Schematics of electronic structure in Weyl semimetals. (a) Weyl points with opposite chiralities. The directions and magnitudes of the Berry curvatures are shown by the arrows and their colors respectively. (b) Tilt parameter $|v_2/v_1|$ dependence of the Weyl cone*

5.2.4. *Nodal line magnets*

Weyl points are point nodes in the band structure (zero-dimensional band crossing). Such band crossing problems can be generalized and categorized by considering the dimensions and symmetry (Chiu et al. 2016). For example, the following Hamiltonian can host a one-dimensional band crossing called the nodal line,

$$\mathcal{H}(\boldsymbol{k}) = k_z\sigma_y + \left(a - f\left(k_x, k_y\right) - k_z^2\right)\sigma_z, \quad [5.14]$$

and its eigen energy is

$$\varepsilon_\pm(\boldsymbol{k}) = \pm\sqrt{k_z^2 + \left(a - f\left(k_x, k_y\right) - k_z^2\right)^2}, \quad [5.15]$$

where the band touching occurs at $k_z = 0, f\left(k_x, k_y\right) = a$. Figure 5.3a and b shows the ring-shaped $f\left(k_x, k_y\right) = k_x^2 + k_y^2$ $(a > 0)$ and line-shaped $f\left(k_x, k_y\right) = k_x$ nodal lines at $k_z = 0$ plane, respectively. The nodal line in equation [5.14] can be easily gapped out by introducing a term proportional to σ_x. Thus, the band touching can be topologically stable only if σ_x is prohibited by the symmetry such as *xy* mirror plane. Consequently, many nodal lines appear at high symmetry points. Besides, the nodal ring gaps out at $a < 0$ by itself, in contrast sharply with the Weyl points, where creation/annihilation always occurs in pairs. We note that nodal lines are well defined in the absence of the spin–orbit coupling (SOC) and a finite but small gap $\sim \Delta\varepsilon_{\rm SOC}$ opens along them after introducing the SOC. Thus, such small gap structure enhances the Berry curvature (equation [5.9]) and contributes to the large ANE.

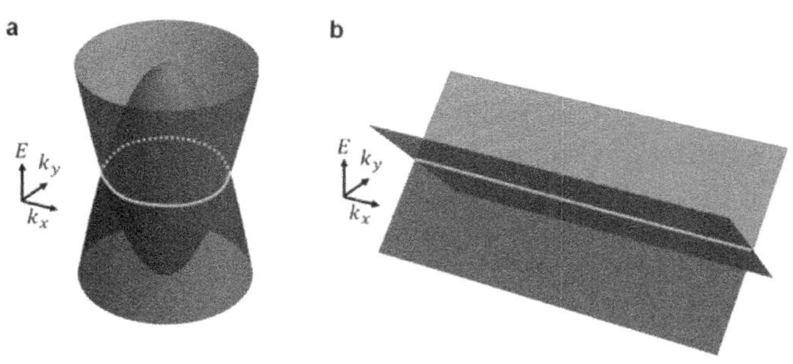

Figure 5.3. *Schematics of the electronic structure near the nodal lines. (a) Ring-shaped nodal line. (b) Line-shaped nodal line*

5.3. Experimental realization of the giant anomalous Nernst effect

5.3.1. *Weyl antiferromagnets Mn_3X (X = Sn, Ge)*

The hexagonal chiral antiferromagnets Mn_3X (X = Sn, Ge) are the first examples of Weyl magnets (Kuroda et al. 2017). Surprisingly, these materials show sizable spontaneous anomalous Hall effects, ANE and magneto-optical Kerr effects in the antichiral 120 degree non-collinear antiferromagnetic phase below $T_N \sim 430$ K for Mn_3Sn and below $T_N \sim 370$ K for Mn_3Ge (Chen et al. 2020), despite the vanishingly small net magnetization (Nakatsuji et al. 2015; Kiyohara et al. 2016; Ikhlas and Tomita 2017; Higo et al. 2018; Chen et al. 2021). Such unique properties, including the first observation of magnetic spin Hall effect (Kimata et al. 2019), and the first electronic manipulation of a Weyl semimetallic state (Tsai et al. 2020) have stimulated the study of antiferromagnetic spintronics (Jungwirth et al. 2016; Šmejkal et al. 2018).

Figure 5.4f shows the magnetization M dependence of the AN coefficient S_{ji}. S_{ji} for Mn_3X (X = Sn, Ge) is highly enhanced and two to three orders of magnitude higher than the values expected for the conventional scaling relations of ferromagnets $S_{ji} \propto M$. However, the magnitude of S_{ji} is only the order of ~0.1 μV/K at room temperature. Nonetheless, the vanishingly small shape anisotropy due to its antiferromagnetism stabilizes the magnetic domain with a large coercivity, and this enables the fabrication of a heat flux sensor robust against the magnetic field perturbations (Higo et al. 2021). Further enhancing S_{ji} in antiferromagnets would be very beneficial for the TE applications by ANE.

5.3.2. *Weyl ferromagnet Co_2MnGa*

The first giant ANE at room temperature was found in the Heusler ferromagnet Co_2MnGa (Figure 5.4a,c,e) (Sakai et al. 2018), whose Curie temperature is $T_C \sim 694$ K (Webster 1971). The Heusler alloys have been extensively studied in the field of spintronics as candidates for the half metal (Varaprasad et al. 2009), and recently, they received more attention owing to the possible topological characters in their electronic structures (Kübler and Felser 2016; Wang et al. 2016). As shown in Figure 5.4e, the AN coefficient S_{ji} reaches ~6 μV/K at room temperature and further increases to ~8 μV/K at 400 K. This value is more than one order of magnitude larger than that for conventional ferromagnets (Figure 5.4f) (Hasegawa et al. 2015). We found that this large ANE originates from the Weyl points near ε_F (Figure 5.4c) by comparing the experimental data to the first-principle calculations. Interestingly, the tilt parameter for these Weyl points is close to the one for the quantum Lifshitz transition $|v_2/v_1| \sim 1$, which enhances both the DOS and the Berry curvature and results in the giant ANE in this material.

The observation of the giant ANE in Co_2MnGa has caused a major impact, leading to various findings such as observation of the Weyl points (Belopolski et al. 2019), giant ANE in a thin-film form (Sumida et al. 2020), anomalous transverse Wiedemann–Franz law (Xu et al. 2020) and giant spin Hall angle (Leiva et al. 2021). Besides, this research also motivates the materials search and leads to the discovery of various ANE materials, including the Fe_3X system mentioned below.

5.3.3. *Nodal-web ferromagnets Fe_3X (X = Ga, Al)*

The discovery of the giant ANE in the Weyl ferromagnet Co_2MnGa have enabled the research and development of new TE conversion based on ANE at room temperature. On the contrary, it is also necessary to improve other factors associated with TE applications such as materials cost and processability. Thus, we searched for new AN materials in iron-based compounds by first performing the high-throughput computational search and then doing real experiments as iron is one of the most inexpensive elements. As a result, we found that the cubic ferromagnets Fe_3Ga and Fe_3Al (Figure 5.4a) exhibit a giant ANE at room temperature. S_{ji} for Fe_3Ga and Fe_3Al, respectively, reaches 6 μV/K and 4 μV/K around room temperature as shown in Figure 5.4e. Based on the first principles calculations, we found that the large S_{ji} for Fe_3X (X = Ga, Al) originates from a nodal-web structure, namely, a flat band structure made of interconnected nodal lines (Figure 5.4d), which also enhance the DOS and the Berry curvature at the same time.

In addition to the giant ANE, the low material cost and high processability for the Fe_3X (X = Ga, Al) system are also beneficial for developing ANE devices especially by using their thin-film form. Indeed, heat flux sensors based on ANE of Fe-Al alloy thin films have been fabricated on thin flexible polyimide sheets (Zhou and Sakuraba 2020). While the sensitivity is still two to three orders of magnitude smaller than that for commercial heat flux sensors based on the Seebeck effect, there are a lot of room for improvement in the film quality and device structure. Further study is necessary for developing ANE-based TE devices.

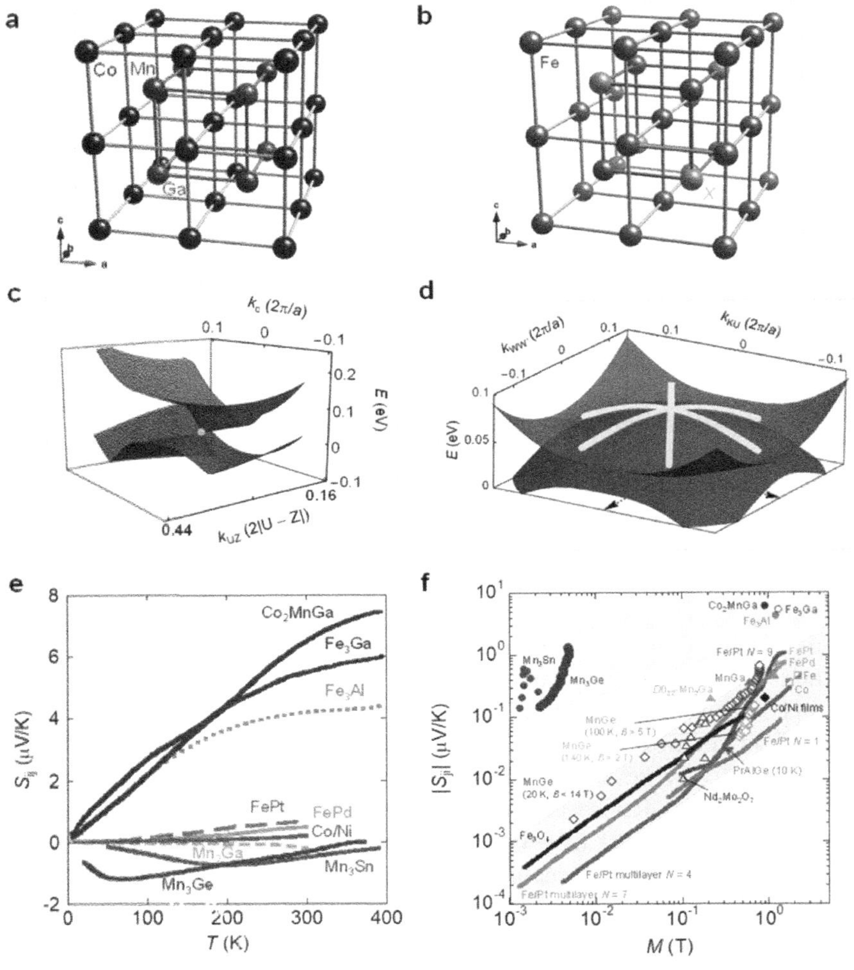

Figure 5.4. *Giant anomalous Nernst effect in topological magnets. (a and b) Crystal structures of Co_2MnGa (a) and Fe_3X (X = Ga, Al) (b). (c and d) Weyl point in Co_2MnGa (c) and nodal web in Fe_3X (d) obtained from the first-principles calculations. Adapted from Sakai et al. (2018, 2020). (e and f) Temperature (e) and magnetization (f) dependences of anomalous Nernst coefficient S_{ji} for Co_2MnGa, Fe_3X (X = Ga, Al), Mn_3X (X = Sn, Ge) and various conventional ferromagnets*

5.4. Summary and prospects

TE generation based on the ANE has various potential benefits and attracts a lot of attention these days owing to the giant ANE found in topological magnets. A number of AN materials have already been found including $UCo_{0.8}Ru_{0.2}Al$ showing "colossal" ANE S_{ji} ~ 23 μV/K at 40 K (Guin et al. 2019; Asaba et al. 2021; He et al. 2021; Nakamura et al. 2021; Chen et al. 2022). Such new findings promote the research for developing AN devices such as heat flux sensors as well as the discovery of novel topological phases in magnets. We anticipate that these studies will lead to innovative TE devices that can be widely used in the future.

5.5. Acknowledgment

This study was conducted in collaboration with Tomoya Higo, Takahiro Tomita, Muhammad Ikhlas, Taishi Chen, Yangming Wang, Susumu Minami, Rieko Ishii, Daisuke Hamane, Pallab Goswami, Michito Suzuki, Takashi Koretsune, Ryotaro Arita and his group members at the University of Tokyo and RIKEN, Yoshichika Otani and his group members at RIKEN and the Institute for Solid State Physics (ISSP), Shinji Miwa and his group members at ISSP, Collin Broholm and his group members at Johns Hopkins University, Chia-Ling Chien and his group members at Johns Hopkins University, Fumiyuki Ishii and his group members at Kanazawa University and A.A. Nugroho and his group members at Bandung Institute of Technology. This study was partially supported by JST-CREST (JPMJCR15Q5, JPMJCR18T3) and Grants-in-Aid for Scientific Research (19H00650). The Institute for Quantum Matter, an Energy Frontier Research Center, was funded by DOE, Office of Science, Basic Energy Sciences under Award DE-SC0019331.

5.6. References

Armitage, N.P., Mele, E.J., Vishwanath, A. (2018). Weyl and Dirac semimetals in three-dimensional solids. *Rev. Mod. Phys.*, 90, 15001.

Asaba, T., Ivanov, V., Thomas, S.M., Savrasov, S.Y., Thompson, J.D., Bauer, E.D., Ronning, F. (2021). Colossal anomalous Nernst effect in a correlated noncentrosymmetric kagome ferromagnet. *Sci. Adv.*, 7(13).

Belopolski, I., Manna, K., Sanchez, D.S., Chang, G., Ernst, B., Yin, J., Zhang, S.S., Cochran, T., Shumiya, N., Zheng, H. et al. (2019). Discovery of topological Weyl fermion lines and drumhead surface states in a room temperature magnet. *Science*, 365, 1278–1281.

Chen, Y., Gaudet, J., Dasgupta, S., Marcus, G.G., Lin, J., Chen, T., Tomita, T., Ikhlas, M., Zhao, Y., Chen, W.C. et al. (2020). Antichiral spin order, its soft modes, and their hybridization with phonons in the topological semimetal Mn_3Ge. *Phys. Rev. B*, 102, 054403.

Chen, T., Tomita, T., Minami, S., Fu, M., Koretsune, T., Kitatani, M., Muhammad, I., Nishio-Hamane, D., Ishii, R., Ishii, F. et al. (2021). Anomalous transport due to Weyl fermions in the chiral antiferromagnets Mn_3X, X = Sn, Ge. *Nat. Commun.*, 12, 1–14.

Chen, T., Minami, S., Sakai, A., Wang, Y., Feng, Z., Nomoto, T., Hirayama, M., Ishii, R., Koretsune, T., Arita, R., Nakatsuji, S. (2022). Large anomalous Nernst effect and nodal plane in an iron-based kagome ferromagnet. *Sci. Adv.*, 8, eabk1480.

Chiu, C.-K., Teo, J.C.Y., Schnyder, A.P., Ryu, S. (2016). Classification of topological quantum matter with symmetries. *Rev. Mod. Phys.*, 88, 035005.

Guin, S.N., Vir, P., Zhang, Y., Kumar, N., Watzman, S.J., Fu, C., Liu, E., Manna, K., Schnelle, W., Gooth, J. et al. (2019). Zero-field Nernst effect in a ferromagnetic kagome-lattice Weyl-semimetal $Co_3Sn_2S_2$. *Adv. Mater.*, 31, 1806622.

Hasegawa, K., Mizuguchi, M., Sakuraba, Y., Kamada, T., Kojima, T., Kubota, T., Mizukami, S., Miyazaki, T., Takanashi, K. (2015). Material dependence of anomalous Nernst effect in perpendicularly magnetized ordered-alloy thin films. *Appl. Phys. Lett.*, 106, 252405.

He, B., Şahin, C., Boona, S.R., Sales, B.C., Pan, Y., Felser, C., Flatté, M.E., Heremans, J.P. (2021). Large magnon-induced anomalous Nernst conductivity in single-crystal MnBi. *Joule*. doi:10.1016/j.joule.2021.08.007.

Higo, T., Man, H., Gopman, D.B., Wu, L., Koretsune, T., van 't Erve, O.M.J., Kabanov, Y.P., Rees, D., Li, Y., Suzuki, M.T. et al. (2018). Large magneto-optical Kerr effect and imaging of magnetic octupole domains in an antiferromagnetic metal. *Nat. Photon.*, 12, 73–78.

Higo, T., Li, Y., Kondou, K., Qu, D., Ikhlas, M., Uesugi, R., Nishio-Hamane, D., Chien, C.L., Otani, Y., Nakatsuji, S. (2021). Omnidirectional control of large electrical output in a topological antiferromagnet. *Adv. Funct. Mater.*, 31, 2008971.

Ikhlas, M., Tomita, T., Koretsune, T., Suzuki, M.T., Nishio-Hamane, D., Arita, R., Otani, Y., Nakatsuji, S. (2017). Large anomalous Nernst effect at room temperature in a chiral antiferromagnet. *Nat. Phys.*, 13, 1085–1090.

Jungwirth, T., Marti, X., Wadley, P., Wunderlich, J. (2016). Antiferromagnetic spintronics. *Nat. Nanotechnol.*, 11, 231–241.

Kimata, M., Chen, H., Kondou, K., Sugimoto, S., Muduli, P.K., Ikhlas, M., Omori, Y., Tomita, T., MacDonald, A.H., Nakatsuji, S. et al. (2019). Magnetic and magnetic inverse spin Hall effects in a non-collinear antiferromagnet. *Nature*, 565, 627–630.

Kiyohara, N., Tomita, T., Nakatsuji, S. (2016). Giant anomalous Hall effect in the chiral antiferromagnet Mn_3Ge. *Phys. Rev. Appl.*, 5, 064009.

Kübler, J. and Felser, C. (2016). Weyl points in the ferromagnetic Heusler compound Co_2MnAl. *Epl.*, 114, 47005.

Kuroda, K., Tomita, T., Suzuki, M.-T., Bareille, C., Nugroho, A.A., Goswami, P., Ochi, M., Ikhlas, M., Nakayama, M., Akebi, S. et al. (2017). Evidence for magnetic Weyl fermions in a correlated metal. *Nat. Mater.*, 16, 1090–1095.

Leiva, L., Granville, S., Zhang, Y., Dushenko, S., Shigematsu, E., Shinjo, T., Ohshima, R., Ando, Y., Shiraishi, M. (2021). Giant spin hall angle in the heusler alloy Weyl ferromagnet Co_2MnGa. *Phys. Rev. B*, 103, L041114.

Mizuguchi, M. and Nakatsuji, S. (2019). Energy-harvesting materials based on the anomalous Nernst effect. *Sci. Technol. Adv. Mater.*, 20, 262–275.

Nagaosa, N., Sinova, J., Onoda, S., MacDonald, A.H., Ong, N.P. (2010). Anomalous Hall effect. *Rev. Mod. Phys.*, 82, 1539–1592.

Nakamura, H., Minami, S., Tomita, T., Nugroho, A.A., Nakatsuji, S. (2021). Logarithmic criticality in transverse thermoelectric conductivity of the ferromagnetic topological semimetal CoMnSb. *Phys. Rev. B*, 104.

Nakatsuji, S. and Arita, R. (2022). Topological magnets: Functions based on Berry phase and multipoles. *Annu. Rev. Condens. Matter Phys.*, 13, 119–142.

Nakatsuji, S., Kiyohara, N., Higo, T. (2015). Large anomalous Hall effect in a non-collinear antiferromagnet at room temperature. *Nature*, 527, 212–215.

Nielsen, H.B. and Ninomiya, M. (1981). Absence of neutrinos on a lattice: (I). Proof by homotopy theory. *Nucl. Phys. B*, 185, 20–40.

Nolas, G.S., Sharp, J., Goldsmid, J. (2001). *Thermoelectrics: Basic Principles and New Materials Developments*. Springer, London.

Sakai, A., Mizuta, Y.P., Nugroho, A.A., Sihombing, R., Koretsune, T., Suzuki, M.T., Takemori, N., Ishii, R., Nishio-Hamane, D., Arita, R. et al. (2018). Giant anomalous Nernst effect and quantum-critical scaling in a ferromagnetic semimetal. *Nat. Phys.*, 14, 1119–1124.

Sakai, A., Minami, S., Koretsune, T., Chen, T., Higo, T., Wang, Y., Nomoto, T., Hirayama, M., Miwa, S., Nishio-Hamane, D. et al. (2020). Iron-based binary ferromagnets for transverse thermoelectric conversion. *Nature*, 581, 53–57.

Sakuraba, Y., Hasegawa, K., Mizuguchi, M., Kubota, T., Mizukami, S., Miyazaki, T., Takanashi, K. (2013). Anomalous Nernst effect in $L1_0$-FePt/MnGa thermopiles for new thermoelectric applications. *Appl. Phys. Express*, 6, 0–4.

Šmejkal, L., Mokrousov, Y., Yan, B., MacDonald, A.H. (2018). Topological antiferromagnetic spintronics. *Nat. Phys.*, 14, 242–251.

Soluyanov, A.A., Gresch, D., Wang, Z., Wu, Q., Troyer, M., Dai, X., Bernevig, B.A. (2015). Type-II Weyl semimetals. *Nature*, 527, 495–498.

Sumida, K., Sakuraba, Y., Masuda, K., Kono, T., Kakoki, M., Goto, K., Zhou, W., Miyamoto, K., Miura, Y., Okuda, T. et al. (2020). Spin-polarized Weyl cones and giant anomalous Nernst effect in ferromagnetic Heusler films. *Commun. Mater.*, 1, 1–9.

Thouless, D.J., Kohmoto, M., Nightingale, M.P., den Nijs, M. (1982). Quantized Hall conductance in a two-dimensional periodic potential. *Phys. Rev. Lett.*, 49, 405.

Tsai, H., Higo, T., Kondou, K., Nomoto, T., Sakai, A., Kobayashi, A., Nakano, T., Yakushiji, K., Arita, R., Miwa, S. et al. (2020). Electrical manipulation of a topological antiferromagnetic state. *Nature*, 580, 608–613.

Varaprasad, B.S.D.C.S., Rajanikanth, A., Takahashi, Y.K., Hono, K. (2009). Highly spin-polarized $Co_2MnGa_{0.5}Sn_{0.5}$ Heusler compound. *Acta Mater.*, 57, 2702–2709.

Wan, X., Turner, A.M., Vishwanath, A., Savrasov, S.Y. (2011). Topological semimetal and Fermi-arc surface states in the electronic structure of pyrochlore iridates. *Phys. Rev. B – Condens. Matter Mater. Phys.*, 83, 1–9.

Wang, Z., Vergniory, M.G., Kushwaha, S., Hirschberger, M., Chulkov, E.V., Ernst, A., Ong, N.P., Cava, R.J., Bernevig, B.A. (2016). Time-reversal-breaking Weyl fermions in magnetic Heusler alloys. *Phys. Rev. Lett.*, 117, 236401.

Webster, P.J. (1971). Magnetic and chemical order in Heusler alloys containing cobalt and manganese. *J. Phys. Chem. Solids*, 32, 1221–1231.

Xiao, D., Chang, M.-C., Niu, Q. (2010). Berry phase effects on electronic properties. *Rev. Mod. Phys.*, 82, 1959.

Xu, L., Li, X., Ding, L., Chen, T., Sakai, A., Fauqué, B., Nakatsuji, S., Zhu, Z., Behnia, K. (2020). Anomalous transverse response of Co_2MnGa and universality of the room-temperature αijA/σijA ratio across topological magnets. *Phys. Rev. B*, 101, 180404.

Zhou, W. and Sakuraba, Y. (2020). Heat flux sensing by anomalous Nernst effect in Fe–Al thin films on a flexible substrate. *Appl. Phys. Express*, 13, 043001.

6

A Comprehensive Review of Phonon Engineering

Bin XU, Harsh CHANDRA, and Junichiro SHIOMI
University of Tokyo, Japan

6.1. Introduction

In recent years, climate and environmental problems have boosted the rapid development of technological innovation for achieving carbon neutrality goals. A great deal of effort has been placed on using clean energy and improving energy efficiency. In addition to the progress made on enhancing the energy utilization rate, effectively reusing the large portion of energy that has dissipated in the form of waste heat is essential. Thermoelectrics have a wide range of applications among thermal energy harvesting techniques due to their simple structure and maintenance-free properties. The performance of energy harvesting of thermoelectric materials is indicated by the figure of merit expressed as ZT:

$$ZT = \frac{S^2\sigma}{\kappa}T \qquad [6.1]$$

where T is the temperature, S is the Seebeck coefficient, σ is the electrical conductance and κ is the thermal conductivity. According to this equation [6.1], selectively reducing κ while maintaining high electrical conductivity is the key to achieving high ZT. Here, the reduction of κ is usually achieved by modulating the carriers of lattice thermal energy, i.e. phonons, through

Thermoelectric Micro/Nano Generators 2,
coordinated by Hiroyuki AKINAGA, Atsuko KOSUGA,
Takao MORI and Gustavo ARDILA.

"phonon engineering". Notably, although electrons contribute to thermal conduction in metallic and highly doped semiconductor materials, we focus on reducing the thermal conduction by phonons, as thermoelectric materials require σ to be maintained.

Here is a recent representative example of improving the performance of thermoelectrical materials through phonon engineering. Silicon is the most commonly used semiconductor, with a low price and matured fabricating technique. Despite the high Seebeck coefficient and the highly controllable electrical properties, single-crystal silicon is not considered to be an effective thermoelectric material because of the high k (145 W/m-K at room temperature) and the resulting low ZT value of approximately 0.01. This problem is overcome by introducing a nanostructure, including a nanowires structure, high porosity and small grain size, which decrease the k of silicon to a few W/m-K, promoting the ZT to 0.3 at room temperature (Kashiwagi et al. 2019; Yang et al. 2021).

Phonon engineering not only gives us opportunities to broaden the range of thermoelectric material candidates to those more abundant, less expensive and with more mature manufacturing infrastructures, but can also further improve the energy conversion efficiency of those commercial thermoelectric materials. Poudel et al. (2008) elevated the room temperature ZT of BiSbTe to 1.2 by nanocrystalline through a ball-milling/hot press method. Yan et al. (2012) improved the ZT of half-Heuslers $Hf_{1-x}Zr_xCoSb_{0.8}Sn_{0.2}$ when Zr is replaced by Ti, introducing a significant atomic mass difference between Hf and Ti, resulting in low thermal conductivity. However, the computational method can also help us to predict and explore low thermal conductive lattice structures with excellent thermoelectric performance such as $PbTe_{1-x}Se_x$ (Tian et al. 2012) and SnSe (Zhao et al. 2014).

Regarding the importance of phonon engineering for developing thermoelectric materials and the enormous progress that has been made recently, it is essential to have a comprehensive understanding of phonon engineering. This chapter starts by introducing the basic concept of thermal conduction and phonons (section 6.1). Then, we introduce the theoretical calculations and experimental methods for predicting/evaluating thermal conductivity and phonon properties (section 6.2). The next section introduces the mechanism and approaches for phonon engineering and related research advances from both mechanistic and methodological viewpoints (section 6.3). Finally, the chapter concludes with a summary and future prospects.

6.1.1. *Thermal conductivity*

Electrons or atomic vibrations mainly carry heat in solids. The lattice vibrational waves contribute to thermal conductivity predominantly in a dielectric crystalline material and majorly in many semiconductors. In principle, thermal conductivity represents how easily a material can conduct heat across a given temperature gradient per unit cross-sectional area of a material. The thermal conductivity is defined through Fourier's law of heat conduction:

$$q = -\kappa \nabla T \quad [6.2]$$

where q is the local heat flux (W/m^2), κ is the lattice thermal conductivity (W/m-K) and ∇T is the temperature gradient (K/m). κ has an extensive range, as illustrated in Figure 6.1, and the materials are used extensively for applications such as thermal insulation and thermal management in electronics, buildings, automobiles and aerospace.

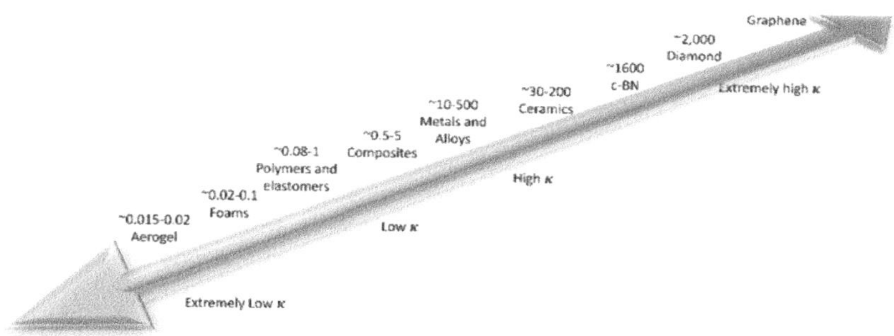

Figure 6.1. *Thermal conductivity (κ in units of W/m-K) of different materials spanning five orders of magnitude*

6.1.2. *Phonons in thermal transport*

6.1.2.1. *Harmonic properties of phonons*

For a crystal semiconductor composed of dielectric material, interatomic potential $V(x)$ can be Taylor-series expanded in infinitesimal displacements from equilibrium $x = r_0$:

$$V(x) = V(r_0) + \left(\frac{\partial V}{\partial r}\right)_0 x + \frac{1}{2}\left(\frac{\partial^2 V}{\partial r^2}\right)_0 x^2 + \ldots \quad [6.3]$$

Since interatomic force is the derivative of potential with respect to the displacement, the lattice dynamics are characterized with the third harmonic term and higher order terms. The harmonic approximation shown in Figure 6.2 that ignores the higher order terms is often used to identify the phonon states, i.e. the eigenvalue and eigenmode of the phonons, although higher order terms need to be considered to account for phonon–phonon scattering, and even phonon states are known to be influenced by higher order terms for crystal with large anharmonicity.

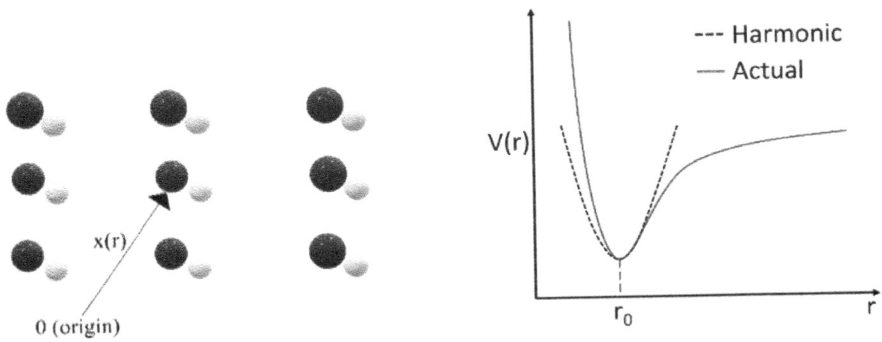

Figure 6.2. *Schematic of crystal lattice on the left side with position vector given by x(r) from the origin (0,0). The right-side figure represents the deviation of the interatomic potential energy of the actual with respect to that given by harmonic approximation, which is parabolic at small displacements around the mean position. For a color version of this figure, see www.iste.co.uk/akinaga/thermoelectric2.zip*

A phonon is described by the vibrational frequency ω and the wave vector $\mathbf{q}$. $\mathbf{q}$ is related to wavelength λ as:

$$\mathbf{q} = \frac{2\pi}{\lambda}\mathbf{q_0} \quad [6.4]$$

where $\boldsymbol{q_0}$ is the unit vector of $\boldsymbol{q}$, along the propagation direction of the wave vector. The propagation velocity of a phonon is given by the group velocity.

$$v_g = \frac{\partial \omega}{\partial q} \quad [6.5]$$

This is obtained from the dispersion relations, the relationship between ω and $\mathbf{q}$, which can be calculated by the dynamical matrix combing equation

[6.3] and the equation of motion. The group velocity tends to the sound speed when $\boldsymbol{q} \rightarrow \mathbf{0}$ in the limiting case that wavelengths are large.

To introduce the lattice dynamics, we begin with the one-dimensional (1D) atom chain (see Figure 6.3). From Newton's Second Law of motion and Hooke's Law, the equation of motion of the nth atom in the monoatomic chain is given by

$$M\frac{d^2x_n}{dt^2} = k[(x_{n+1} - x_n) + (x_{n-1} - x_n)] \qquad [6.6]$$

where x_n is the displacement of the nth atom from the mean equilibrium position, and k is the interatomic nearest neighbor force constant. Here, M is the mass of an atom and k is the nearest-neighbor force constant. We can try a solution $x_n = x_{amp_n}\exp[\mathrm{i}(qna - \omega t)]$ for $x = na$, where x_{amp_n} means the amplitude of vibration of the nth atom.

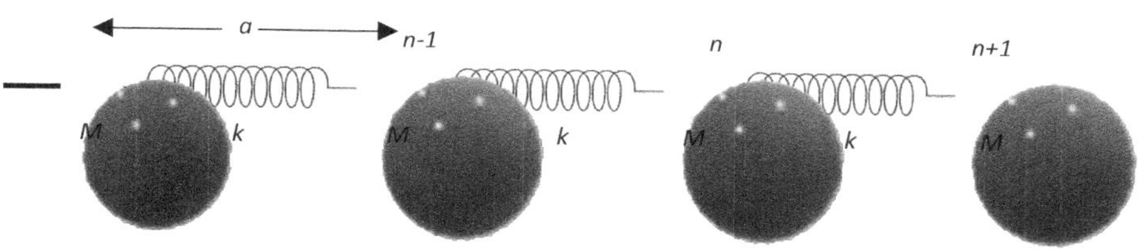

Figure 6.3. *Monoatomic linear chain exhibiting motion from equilibrium position for the (n-1)th, nth and (n+1)th atoms with force constants k*

Solving equation [6.7] yields the following dynamical equation:

$$\omega^2 x_{amp_n} = \frac{2k}{M}(1 - cos(\mathbf{q}a)) \qquad [6.7]$$

Simplifying for ω gives

$$\omega = 2\sqrt{\frac{2k}{M}}\left|sin\left(\frac{qa}{2}\right)\right| \qquad [6.8]$$

This simple phonon dispersion is illustrated in Figure 6.4, with only one acoustic branch.

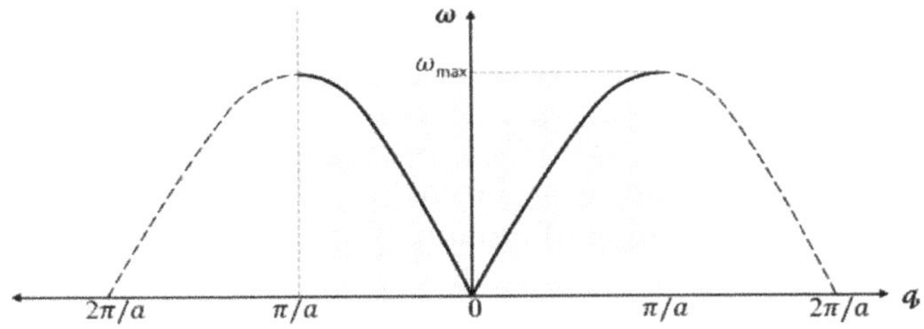

Figure 6.4. *Phonon dispersion curve for monoatomic linear chain*

For a more complex case with a basis of two atoms with masses m and M for a diatomic linear chain, the lattice dynamical is shown in Figure 6.5.

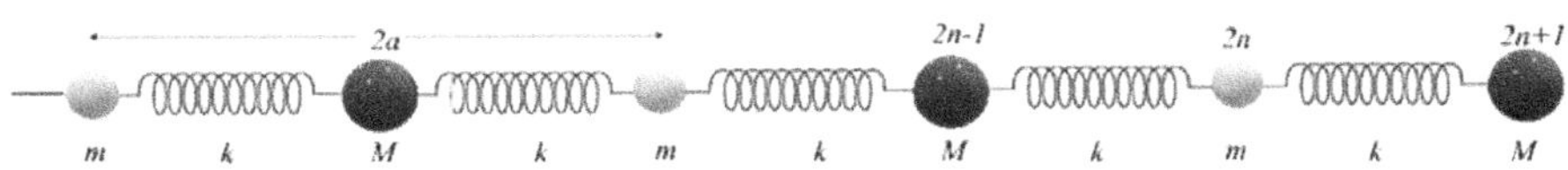

Figure 6.5. *Diatomic linear chain with masses m and M with unit cell length 2a exhibiting displacements from the mean position*

The equation of motion can be written using Newton's Second Law and Hooke's Law as:

$$m\frac{d^2x_{2n}}{dt^2} = k[(x'_{2n+1} - x_{2n}) + (x'_{2n-1} - x_{2n})] \quad [6.9]$$

$$M\frac{d^2x'_{2n}}{dt^2} = k[(-x'_{2n+1} + x_{2n}) + (-x'_{2n-1} + x_{2n})] \quad [6.10]$$

where x' is the displacement of atoms with mass M from their equilibrium position, and x is the displacement of atoms with mass m from their equilibrium position.

Moving ahead, we will determine the solution of x_{2n} and x_{2n+1}. Solving equations [6.10] and [6.11] yields the following dynamical equation:

$$\omega^2 x_{amp_i} = \sum_{j=1}^{2} D_{ij}\, x_{amp_i} \quad [6.11]$$

where D is a second-order dynamical matrix given by:

$$D = \begin{pmatrix} 2k/m & -\frac{2k}{M}cos(\mathbf{q}a) \\ -\frac{2k}{M}cos(\mathbf{q}a) & 2k/M \end{pmatrix} \qquad [6.12]$$

Thus, the eigenfrequencies obtained from equation [6.12] can be simplified as follows:

$$\omega^2 = k\left(\frac{1}{m}+\frac{1}{M}\right) \pm \left[\left(\frac{1}{m}+\frac{1}{M}\right)^2 - \frac{4}{mM}\{\sin(qa)\}^2\right]^{1/2} \qquad [6.13]$$

From the obtained ω^2 function, we can see two different kinds of frequencies involved at each wave vector point, which gives the phonon dispersion curve of a diatomic chain (Figure 6.6).

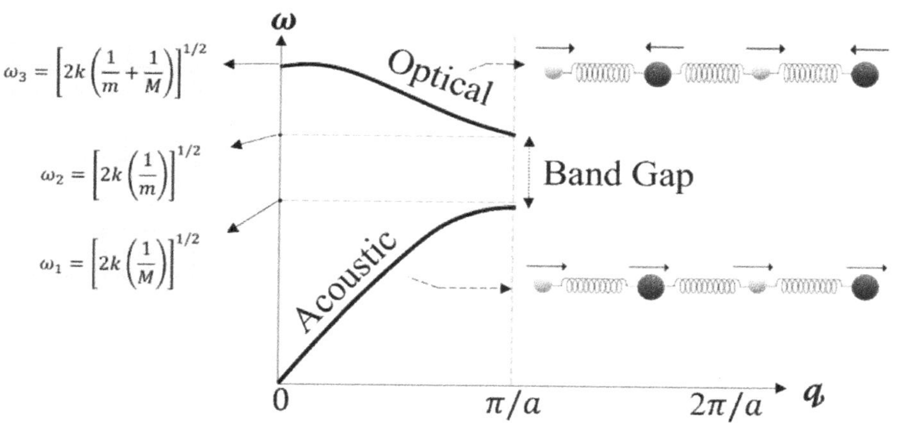

Figure 6.6. *Phonon dispersion shows two branches due to two masses m and M in the chain*

In Figure 6.6, the branch with a higher frequency is called the optical phonon branch, resulting from atoms in a unit cell moving out of phase directions or in opposite directions. The other branch with lower frequency is the "acoustic branch", which results from atoms moving in-phase with each other, as in the case of the monoatomic linear chain (Kittel et al. 1996).

6.1.2.2. *Anharmonic properties of phonons*

Consider a crystal with b atoms per unit cell, and let $u(ap)$ represent the displacement of the pth atom in the ath unit cell. Assume that the total potential energy $\mathcal{V}$ of the crystal is a function of the instantaneous position of all atoms. We can now expand $\mathcal{V}$ in a Taylor series in powers of the atomic displacements (Ashcroft and Mermin 1976):

$$\begin{aligned}\mathcal{V} &= \mathcal{V}_0 + \sum_{ap\alpha} \left.\frac{\partial \mathcal{V}}{\partial u_\alpha(ap)}\right|_0 u_\alpha(ap) \\ &+\frac{1}{2}\sum_{ap,a'p}\sum_{\alpha\beta} \left.\frac{\partial^2 \mathcal{V}}{\partial u_\alpha(ap)\partial u_\beta(a'p)}\right|_0 u_\alpha(ap)u_\beta(a'p') \\ &\quad+\frac{1}{3!}\sum_{ap,a'p',a''p''}\sum_{\alpha\beta\gamma} \left.\frac{\partial^3 \mathcal{V}}{\partial u_\alpha(ap)\partial u_\beta(a'p')\partial u_\gamma(a''p'')}\right|_0 \\ &\qquad\times u_\alpha(ap)u_\beta(a'p')u_\gamma(a''p'') + \cdots \\ &= \mathcal{V}_0 + \mathcal{V}_1 + \mathcal{V}_2 + \mathcal{V}_3 + \cdots\end{aligned} \quad [6.14]$$

where $u_\alpha(ap)$ refers to atom (ap) along the α direction, $u_\beta(a'p')$ refers to atom $(a'p')$ along the $\beta\ direction$, and $u_\gamma(a''p'')$ refers to atom (ap) along the γ direction.

At equilibrium, the second term in the above expansion vanishes because the first derivative of potential energy at the mean position is 0. Moreover, the first term can be set to 0 as it implies zero potential. Therefore, we are left with terms starting from the second and beyond as rewritten below:

$$\begin{aligned}\mathcal{V} &= \frac{1}{2}\sum_{ap,a'p}\sum_{\alpha\beta} \Theta_{\alpha\beta}(ap,a'p')u_\alpha(ap)u_\beta(a'p') \\ &+\frac{1}{3!}\sum_{ap,a'p',a''p''}\sum_{\alpha\beta\gamma} \Psi_{\alpha\beta\gamma}(ap,a'p',a''p'') \\ &\quad\times u_\alpha(ap)u_\beta(a'p')u_\gamma(a''p'') + \cdots\end{aligned} \quad [6.15]$$

where Θ and Ψ are tensors of the second and third rank, expanded as:

$$\Theta_{\alpha\beta}(ap,a'p') = \frac{\partial^2 \mathcal{V}}{\partial u_\alpha(ap)\partial u_\beta(a'p)} \quad [6.16]$$

$$\Psi_{\alpha\beta\gamma}(ap,a'p',a''p'') = \frac{\partial^3 \mathcal{V}}{\partial u_\alpha(ap)\partial u_\beta(a'p')\partial u_\gamma(a''p'')} \quad [6.17]$$

The introduction of the third term $\Psi_{\alpha\beta\gamma}$ in the interatomic potential function results in interaction between phonons, which are otherwise discussed independently in the anharmonic theoretical treatment of phonons. The third term produces the first-order perturbation, which refers to the three phonon scattering, while the following high-order perturbation is contributed by more complicated interactions engaged by more phonons. The perturbation terms result in a deviation from the harmonic behavior of atomic displacements and are therefore referred to as anharmonicity (shown in Figure 6.7). The anharmonicity is the cause of the finite thermal conductivity of single crystals.

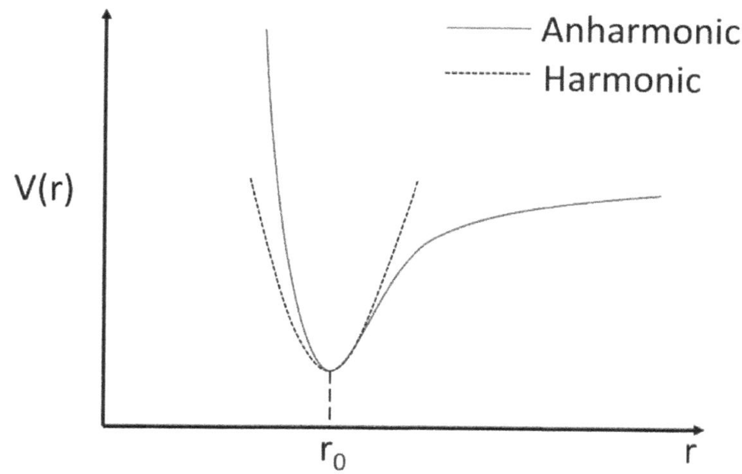

Figure 6.7. *The atomic potential for the actual case and those under the harmonic and anharmonic approximation. For a color version of this figure, see www.iste.co.uk/akinaga/thermoelectric2.zip*

6.1.2.3. *Phonons scattering and transport*

For the case of a three-phonon interaction arising due to the first-order perturbation term of anharmonicity, three involved wave vectors $\mathbf{q}$, $\mathbf{q}_1$ and $\mathbf{q}_2$ have to satisfy momentum conservation given by

$$\mathbf{q} + \mathbf{q}_1 + \mathbf{q}_2 = \mathbf{G} \qquad [6.18]$$

where $\mathbf{G}$ is a reciprocal lattice vector, and $\mathbf{q}$, $\mathbf{q}_1$ and $\mathbf{q}_2$ are constrained to the first Brillouin zone. Three-phonon interactions for which $|\mathbf{G}| = 0$ and $|\mathbf{G}| \neq 0$ are referred to as Normal (N) and Umklapp (U) processes, respectively. There are two kinds of permissible processes (Figure 6.8):

(i) the annihilation of two phonons and the creation of a third phonon (a *type I* event) and (ii) the annihilation of one phonon and the creation of two phonons (a *type 2* event). The three-phonon processes have to satisfy energy and momentum conservation laws along their polarization dispersion branches (optical or acoustic), according to their types mentioned below:

Type I events:

$$\mathbf{q} + \mathbf{q_1} = \mathbf{q_2} + \mathbf{G} \quad [6.19]$$

$$\omega(\mathbf{q}) + \omega'(\mathbf{q_1}) = \omega''(\mathbf{q_2}) \quad [6.20]$$

Type II events:

$$\mathbf{q} + \mathbf{G} = \mathbf{q_1} + \mathbf{q_2} \quad [6.21]$$

$$\omega(\mathbf{q}) = \omega'(\mathbf{q_1}) + \omega''(\mathbf{q_2}) \quad [6.22]$$

In a U-process, the quasi-momentum of newly generated phonon $\mathbf{q_2}$ differs from incoming phonons $\mathbf{q}$ and $\mathbf{q_1}$ by a non-zero reciprocal lattice vector **G** because **G** flips the resultant net vector, and hence, energy direction back to the first Brillouin zone. For this reason, N- and U-processes are momentum conserving and momentum non-conserving processes, respectively.

There are various approaches that can be employed to formulate the microscopic expression of the thermal transport. One is the Phonon Boltzmann transport equation (BTE), which solves for a distribution function of phonons at a particular location at a particular time in the crystal. Another is the Green's function-based approach, which deals with quantum statistics of disturbances.

We use the relaxation time approximation, where a phonon experiences a collision with a probability per unit time $\frac{1}{\tau}$, i.e. the probability of a phonon undergoing a collision in an infinitesimal time dt is the time interval per unit time $\frac{dt}{\tau}$. Here, τ is called the relaxation time or lifetime.

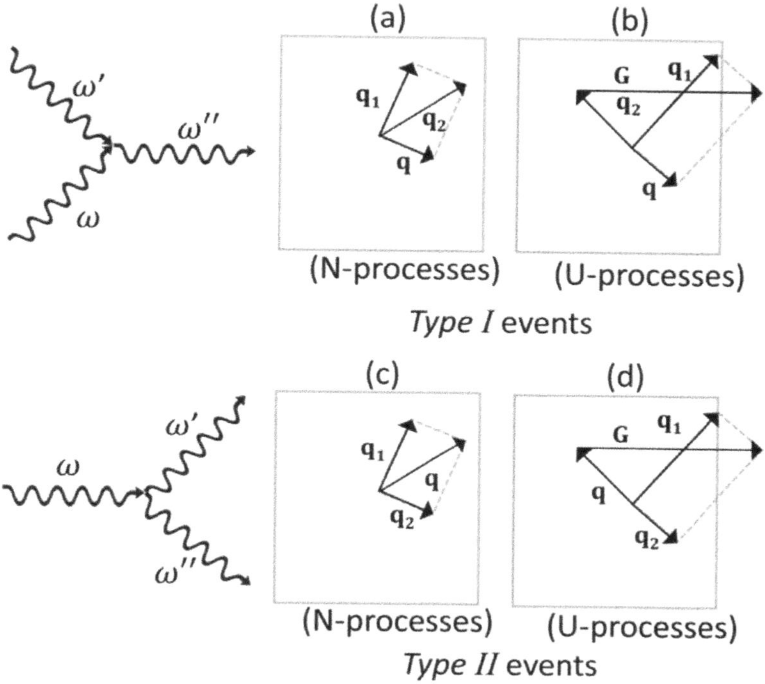

Figure 6.8. *Scattering processes illustrating three-phonon creation and annihilation processes in Type I and Type II events*

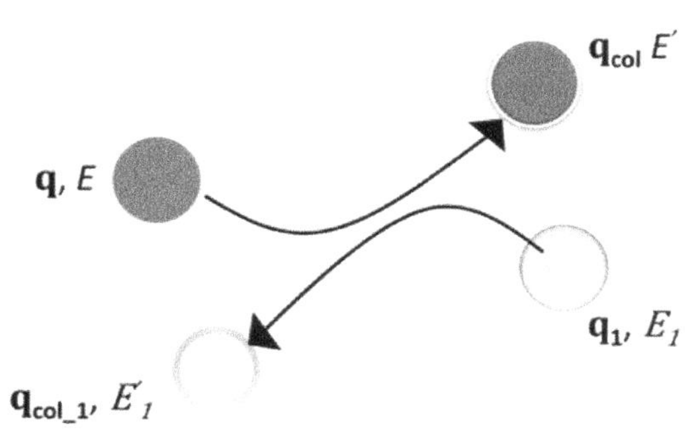

Figure 6.9. *Collision of two carriers with initial wave vector and energy* $(\boldsymbol{q}, E)$ *and* $(\boldsymbol{q}_1, E_1)$. *After the collision, the two particles are at states* $(\boldsymbol{q}_{col}, E'')$ *and* $(\boldsymbol{q}_{col\_1}, \boldsymbol{E}_1')$

In the model, phonons are assumed to achieve thermal equilibrium with their surroundings only through scattering (collision) mechanisms. These collisions are assumed to maintain local thermodynamic equilibrium. A phonon of polarization **b** with wave vector **q** will, as a result of a collision, be scattered into polarization **b** with wave vector $\mathbf{q}_{\mathbf{col}}$, as shown in Figure 6.9.

The lattice thermal conductivity described by BTE can be written as:

$$\kappa = \frac{1}{3}\int \tau v_g{}^2 C(\omega)d\omega = \frac{1}{3}\int_0^{\omega_m} C(\omega)v_g(\omega)\Lambda(\omega)d\omega \qquad [6.23]$$

where $C(\omega)$ is the specific heat per unit frequency at frequency ω.

Based on the Matthiessen rule, and considering the following phonon scattering rates: (1) boundary scattering γ_{bd}, (2) impurity/defect scattering γ_{imp}, and (3) phonon–phonon scattering $\gamma_{\mathrm{ph-ph}}$, τ is given as:

$$\frac{1}{\tau} = \gamma_{\mathrm{bd}} + \gamma_{\mathrm{imp}} + \gamma_{\mathrm{ph-ph}} \qquad [6.24]$$

6.2. Methodology of phonon engineering

6.2.1. *Computational method for thermal conduction and phonon properties*

The thermal energy carried by phonons can be described by the BTE based on the phonon gas model (PGM). The specific heat $C(\omega)$ and group velocity $v(\omega)$ can be calculated by solving the harmonic lattice dynamic. The challenge lies in obtaining the relaxation time of phonon scattering governed by the anharmonic phonon interaction. Researchers compromised on this issue in earlier studies by fitting the experimental result to a simplified dispersion relation, such as the Debye model (Klemens 1955). This approach can explain some experimental results, but the lack of anharmonic details makes the active prediction of the thermal conductivity of new materials almost impossible.

This situation started to change in the late 1990s, when the density functional theory (DFT) method was well developed for calculating the anharmonicity of phonon scattering and enabled the thermal conductivity of new crystal structures to be predicted accurately. Broido et al. (2007) performed the density functional perturbation theory (DFPT) calculation to obtain the thermal conductivity of Si, which is ideally consistent with the experimental result (Broido et al. 2007). A typical procedure for obtaining the thermal conductivity is as follows. First, harmonic and anharmonic interatomic force constants are obtained. This can be done either directly by DFPT or using other numerical methods, such as the displacement method (Esfarjani et al. 2011) and finite difference method (Tian et al. 2012). Second, they are used to perform the harmonic and anharmonic lattice dynamics explained above to calculate the phonon dispersion relation and relaxation time properties. The thermal conductivity κ can then be given by substituting the harmonic properties $C(\omega)$, $v(\omega)$ and anharmonic properties $\tau(\omega)$ into the BTE [6.24].

Not only for calculating the thermal conductivity of bulk crystal, since the thermal conductivity is given as a function of phonon frequency, or the relaxation time τ, the accumulated thermal conductivity to frequency ω or MPF (Λ) can be obtained and provide insight into both the predominated phonon for heat conduction and the "size effect" issue in the Casimir–Knudsen regime.

Despite the incredible power of the DFT calculation in precisely calculating the intrinsic thermal conductivity and details of phonon properties, owing to the complexity in solving the wave equation with high-order force constants, the computational cost is relatively high, especially for large-scale materials with ultra-fine structures such as those with multiple interfaces, embedded with nanoparticles or impurities. Although the accumulated thermal conductivity can be used to predict the size effect of thermal conductivity, since the classical model describing the interfacial phonon transport like acoustic mismatch model (AMM) (Peterson and Anderson 1972) and diffuse mismatch model (DMM) (Swartz and Pohl 1989) do not take the interface details, like the roughness, defects/impurities and the adhesion strength, into account, the actual interfacial thermal conductance would deviate from the prediction by the classical model, and hence results in a significant deviation in the overall thermal conductivity.

Molecular dynamics (MD) simulation is a useful alternative tool to deal with such issues. The classical MD simulation is based on the classical Newton's laws of motion. The principle of MD calculation is much simpler than the first-principles lattice dynamics calculations. By inputting each atom's initial parameters, including the position, potential and velocity, we can predict the motion of individual atoms after a short time interval. By conducting such calculations to all the atoms in the system in cycles, the molecular dynamics of the system can be obtained. Both equilibrium molecular dynamics (EMD) (McGaughey and Larkin 2014) and non-equilibrium molecular dynamics (NEMD) (Shiomi (2014)) simulations can then be performed based on the system to obtain the thermal conductivity. Other important parameters in phonon transport, such as the vibrational density of state (vDOS) and the accumulated thermal conductivity in the frequency regime, can also be calculated by Fourier transformation. MD calculation enables the analysis of phonon transport in a complex system and gives a more accurate prediction dealing with the "size effect" issue. Since MD simulation strongly relies on interatomic potential functions, the key to precise thermal conductivity calculation is to describe the interatomic potential accurately. The progress in the MD simulation methods, such as the ones with machine-learning potential (Zhang et al. 2018), enables a more precise prediction of the thermal conductivity and broadens the range of materials available for MD simulation, regardless of the lack of empirical interatomic potential.

Overall, the theoretical calculation technique not only enables us to predict the thermal conductivity of novel materials and nanostructures, but also provides massive information about phonon behaviors. The theoretical calculation brings a better understanding of phonon engineering and a clear guideline for innovating thermoelectric materials.

6.2.2. *Experimental method for nano-/micro-scale heat conduction characterization*

In phonon engineering, measuring the targeted thermal transport properties at multiple scales is essential. There are many thermal measurement methods for materials at different scales, such as the 3ω method, laser flash, thermoreflectance method and steady-state method. Herein, we mainly focus on the measurement at a small scale, which is typically applied to phonon engineering in current state-of-the-art research works.

6.2.2.1. *Thermoreflectance method*

The ultrafast thermoreflectance pump-probe-based transient time-domain thermoreflectance (TDTR) technique has emerged and evolved extensively for characterizing the thermal conductivity, thermal boundary conductance in nanostructures (Cahill 2004). In TDTR experimental setup, as shown in Figure 6.10, tunable laser excitation sources such as Ti:sapphire laser oscillators with mode-locked function are typically used as the light source. Mode locking enables the tuning of laser wavelength by an extensive range from 700 nm to well above far infrared 1100 nm. The laser source generates a femtosecond width laser pulse-train with a repetition rate of 80 MHz. A broadband Faraday optical isolator is placed at the exit opening of the Ti:sapphire laser oscillator to only allow transmission of the optical signal along one direction and also simultaneously prevent the laser beam from reflecting into the oscillator. A half-wave plate is installed before the broadband Faraday optical isolator to tune the laser power. The incoming laser beam is then split into two beams: a pump beam and a probe beam through a diffractive beam splitter called a polarizing beam splitter (PBS). This results in the pump and the probe beams being cross-polarized to each other. Another half-wave plate before the polarizing beam splitter can be used to tune the power ratio between the pump and the probe beams. The pump beam is modulated at a frequency in the range of 0.1–20 MHz before being directed onto the sample through an objective lens using an EOM (electro-optic modulator). The EOM modulation frequency serves as the reference for lock-in detection because such a lock-in effect makes the system capable of picking up/extracting output from the signal in terms of frequency and amplitude, even though the signal is buried in noise. Next, the probe beam is delayed with respect to the pump beam via a mechanical delay stage, which plays the role of changing the optical path length of the probe beam. The time-delayed probe beam is then directed on the sample, with the probe beam reflecting off the surface of the sample and being detected by a photodiode detector. The detected signal is then processed by the lock-in amplifier.

The lock-in amplifier detects the in-phase voltage signal $V_{inphase} = Re[\Delta\mathbb{R}(t)]$ and out-of-phase voltage signal $V_{out\text{-}of\text{-}phase} = Im[\Delta\mathbb{R}(t)]$, which is essentially the real and imaginary part of $\Delta\mathbb{R}(t)$. Typically, either the amplitude signal $\sqrt{V^2_{\text{out-of-phase}} + V^2_{\text{inphase}}}$ or the ratio signal $V_{\text{inphase}}/V_{\text{out-of-phase}}$ is used in the following fitting process to obtain the targeted parameters. Next, we can subtract the targeted parameters by fitting the decay profile. Figure 6.11 depicts a schematic of the sample structure used in a typical TDTR experimental setup. The thermal modeling of TDTR experiments consists of two steps: the first is to solve three-dimensional anisotropic heat diffusion in a sample made of multilayers with different thermal properties in each stack, and the second is to model the data collected from the experimental measurements theoretically.

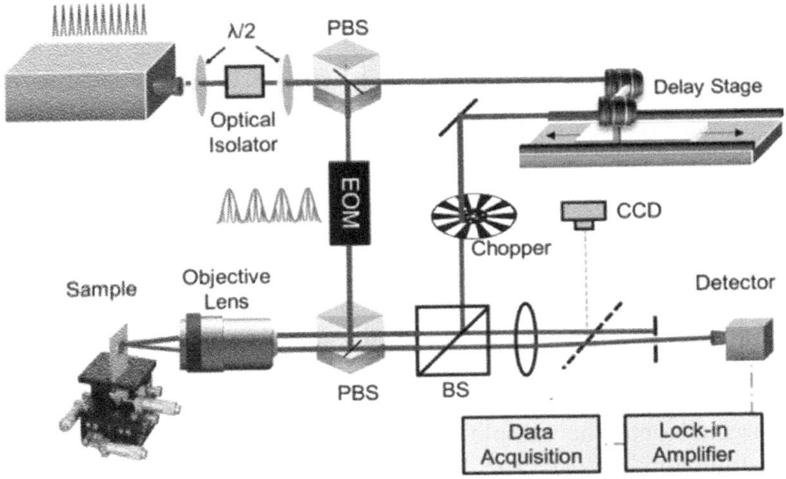

Figure 6.10. *Schematic of femtosecond pump-probe TDTR setup. This figure has been redrawn from Jiang et al. (2018). For a color version of this figure, see www.iste.co.uk/akinaga/thermoelectric2.zip*

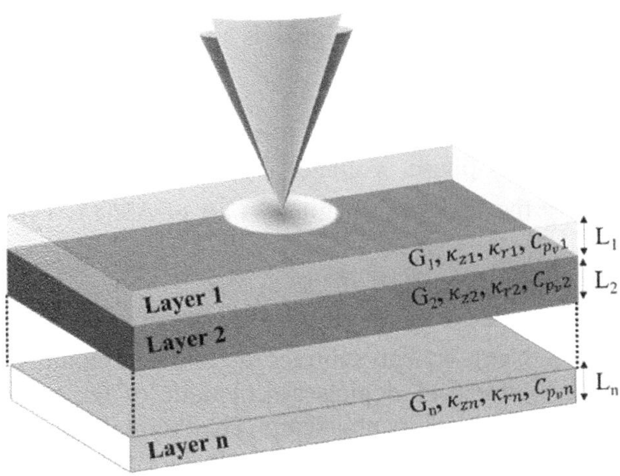

Figure 6.11. *Schematic of a multi-layered stacking of sample, with the top layer representing the transducer layer, followed by sample layers to the end layer representing substrate. Such a sample configuration is essential for the thermal modeling of TDTR experiments for solving the 3D heat diffusion equation*

The governing equation is based on Fourier's rule of heat conduction:

$$C_{p_v}\frac{\partial T}{\partial t} = \frac{\eta\kappa_r}{r}\frac{\partial}{\partial r}\left(r\frac{\partial T}{\partial r}\right) + \kappa_z\frac{\partial^2 T}{\partial z^2} \qquad [6.25]$$

where κ_r and κ_z are the in-plane (along radial direction) and cross-plane (through-plane) thermal conductivity, respectively, and C_{p_v} is the volumetric heat capacity defined as the product of specific heat and density.

6.2.2.2. *3ω method*

The 3ω approach (Cahill 1990), as shown in Figure 6.12, employs a radial heat flow from a single element that serves as both a heater and a thermometer. It is usually used to measure the thermal conductivity of thin film at microscale. In brief, the three frequencies involved in the 3ω approach are the frequency ω (fundamental frequency) corresponding to metal line current, the frequency 2ω corresponding to the temperature variation and the frequency 3ω corresponding to the voltage variation. The third harmonic of the voltage across the line is used to measure the temperature oscillations of the metal line, thus the term 3ω technique. The average temperature fluctuations across the width of the heating line source are given by the following expression when a current is applied to the heater at a frequency ω, i.e. $I(t) = I_0 \cos \omega t$:

$$\Delta T = \frac{P}{l\pi\kappa}\int_0^\infty \frac{\sin^2(kw)}{(kw)^2(k^2+q^2)^{\frac{1}{2}}}dk \qquad [6.26]$$

where w is the finite width of the heater line, q is a complex quantity given by $1/q = (D/i2\omega)^{\frac{1}{2}}$, and k is the Fourier space integration parameter.

If $\ll d_l$, then $\frac{\sin(kw)}{kw} = 1$, and in this case, the ΔT becomes:

$$\begin{aligned}\Delta T &= \frac{P}{\pi\kappa}\left[-\,ln\left(\frac{i2\omega w^2}{D}\right)^{\frac{1}{2}} + \eta\right] \\ &= -\frac{P}{2\pi\kappa}ln\,2\omega - \frac{P}{2\pi\kappa}ln\left(w^2\frac{C_p}{\kappa}\right) \\ &\quad -i\frac{P}{4\kappa} + \frac{\eta P}{\pi\kappa},\end{aligned} \qquad [6.27]$$

where η is a constant with the value of 0.922.

If ≫ d_l , then

$$\lim_{w\to\infty} \frac{1}{\pi w}\left[\frac{\sin wx}{\sin x}\right]^2 = \delta(x) \quad [6.28]$$

ΔT becomes:

$$\Delta T = \frac{\frac{P}{2w}}{\kappa q} = \frac{\frac{P}{2w}}{\sqrt{2\omega C_p \kappa}} e^{-\frac{i\pi}{4}} \quad [6.29]$$

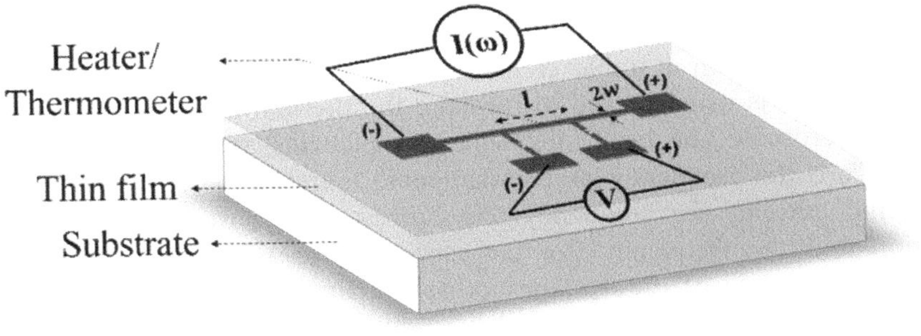

Figure 6.12. *Schematic of 3ω thermal characterization setup, with the metal lines serving as the heater and thermometer atop a thin film (sample) on a substrate*

The aforementioned third harmonic detection approach is then applied to monitor the temperature oscillation of a heater/thermometer itself. Given the temperature coefficient of resistivity of the metal line as $\alpha_c \equiv \left(\frac{1}{R}\right)\left(\frac{dR}{dT}\right)$, the resistance oscillates at the frequency of 2ω is as follows:

$$R(t) = R_0[1 + \alpha \times |\Delta T| \times cos(2\omega t + \phi)] \quad [6.30]$$

The voltage across the heater becomes

$$\begin{aligned} V_{heater}(t) &= I(t)R(t) = I_0 cos\,\omega t \cdot R_0[1 + \alpha \times |\Delta T| \times cos(2\omega t + \phi)] \\ &= I_0 R_0 cos\,\omega t + \tfrac{1}{2} I_0 R_0 \alpha |\Delta T| cos(\omega t + \phi) \\ &+ \tfrac{1}{2} I_0 R_0 \alpha |\Delta T|\, cos(3\omega t + \phi) \end{aligned} \quad [6.31]$$

where ϕ represents the phase shift of ΔT with respect to the power oscillation.

The targeted thermal information of the sample is contained in the second and third components in equation [6.32]. To obtain the last term, which is typical with a small value, a lock-in processing (Moon et al. 1996) is applied using the Wheatstone bridge (see Figure 6.13) to detect the amplitude and phase of the 3ω signal. From the 3ω signal, we can obtain the temperature response and then subtract the targeted thermal conductivity by solving the heat conduction model.

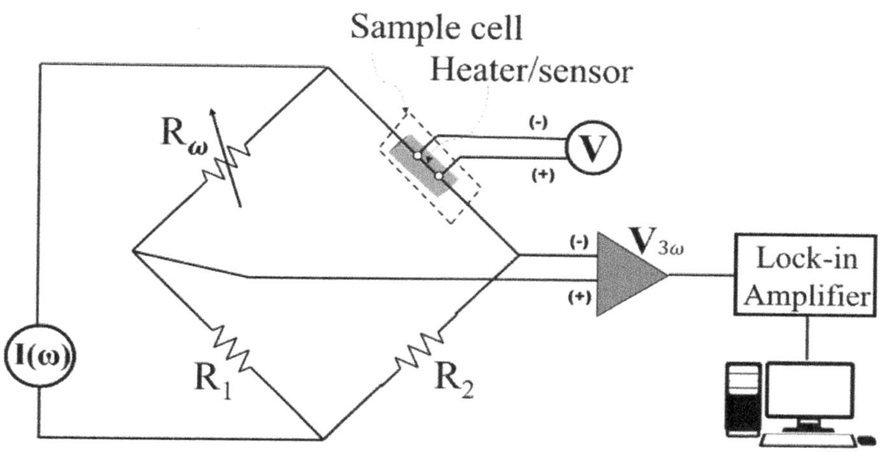

Figure 6.13. *Arrangement for thermal conductivity measurements by a 3-omega method*

6.2.3. *Direct measurement of phonon properties through phonon scattering*

Methods involving phonon scattering with photons, X-rays, electrons and neutrons have been established to probe the phonon properties directly. In this section, we introduce the most commonly used Raman scattering spectroscopy (phonon–photon scattering), neutron scattering and the X-ray scattering method.

6.2.3.1. *Raman scattering method*

When powerful laser beams irradiate the sample in the UV-visible range (v_0) (Ferraro et al. 1994), besides the strong Rayleigh scattering showing the same frequency as the incident beam, there is weak Raman scattering ($\sim 10^{-5}$ of the incident beam) whose frequencies are different from the incident beam. The Raman scattered light manifests in vibrational frequencies that can be blue-shifted $v_0 + v_m$ (Stokes scattering) or red-shifted $v_0 - v_m$ (Anti-stokes scattering) with respect to the original incident beam frequency. The observed Raman shift of the Stokes and anti-Stokes features is a direct measure of the molecule's vibrational energies, since the energy increase or reduction is proportional to the vibrational energy levels in the ground electronic state of the molecule. However, the Raman spectroscopy can only probe long-wavelength/high-frequency phonons confined to gamma point Γ, i.e. to $\boldsymbol{q} \cong 0$ phonons.

Raman scattering is a process in which electromagnetic radiation exchanges vibrating quanta phonons with a crystal through electrons. A thorough understanding of (i) electronic structure, (ii) phonon dispersions and (iii) mutual interactions between electrons and phonons (i.e. electron–electron, electron–phonon and phonon–phonon scattering) are required for a full description. Raman spectroscopy has been an immensely robust and versatile tool to characterize the phonon properties in dielectric materials such as graphite and its derivatives from bulk to monolayer graphene (Basko et al. 2009). The G and D peaks at 1580 cm^{-1} and 1350 cm^{-1} originate from the expansion and contraction modes of six-atom rings and corresponding modes activated by defects, respectively. From the frequency of G peaks, we can obtain the energy of TO phonons at the Brillouin zone's K point.

6.2.3.2. *Neutron and X-ray scattering method*

The primary benefit of neutrons as a probe is to measure fine details of atomic and molecular vibrations via inelastic neutron scattering (INS) (Kohn 1959), which involves the neutron exchanging energy with the atoms in a material. As an example, Kulda et al. (1994) investigated the frequencies and eigenvectors of silicon crystal along the tree main symmetry directions using INS, which shows good agreement with the theoretical calculation. In the mechanism, neutrons can absorb or emit an amount of energy equivalent to a quantum of phonon energy hv, when dispersed by a crystalline material. This results in inelastic scattering, in which the energy of the neutrons before and after scattering differs by a certain amount, $\hbar\omega$ (i.e. energy transfer). Typically, this change is a few terahertz (THz) of frequency in most materials, equivalent to phonon energies of a few meV. As illustrated in Figure 6.14, the neutron has distinctive velocities, and consequently, different wave vectors before and after it interact with the sample. We need to know the neutron wave vector before ($\mathbf{k_i}$) and after ($\mathbf{k_f}$) the scattering event to figure out the phonon energy and scattering vector $\mathbf{Q}$. The neutron can be probed by triple-axis spectrometers (TAS), which can provide both the momentum $\mathbf{Q} = \mathbf{k_i} - \mathbf{k_f}$ and the energy transfer $\hbar\omega = E_i - E_f$. As illustrated in Figure 6.15, the monochromator crystal selects an input neutron beam of a well-defined wave vector $\mathrm{k_I}$ from the neutron source's white spectrum during a TAS experiment (first axis). After that, the monochromatic beam scatters from the crystalline sample (second axis). The analyzer crystal (third axis) reflects the intensity of the dispersed beam with wave vector $\mathrm{k_f}$ onto the neutron detector (^{3}He-tube), estimating the energy transfer $\hbar\omega$.

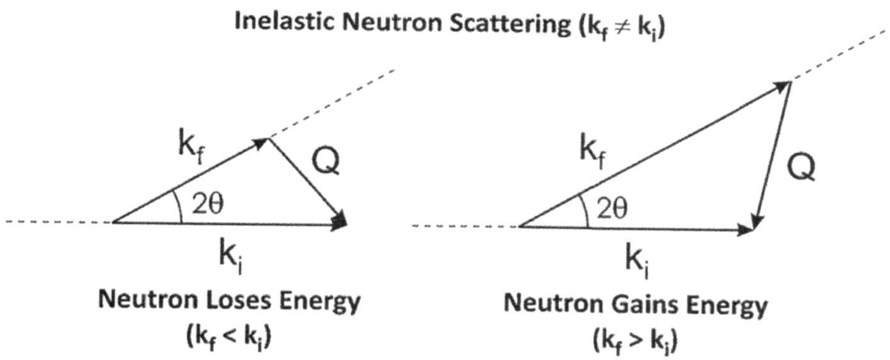

Figure 6.14. *Scattering triangles for inelastic neutron scattering (INS)*

Inelastic X-ray scattering (IXS) is an alternative method for probing the phonon properties based on the inelastic interaction between photon (X-ray) and phonon. Unlike Raman scattering, which is limited to near Γ modes, IXS can probe the phonon dispersion in an extended wave vector regime due to the much smaller wavelength of X-rays, and can provide information about the relaxation time spectrum. Compared with INS, IXS is fit for probing the relaxation time of high-frequency phonons and is beneficial for measuring thin film samples with a thickness of several micrometers using a small incident angle, which enables a relatively small penetration depth. Uchiyama et al. (2018) used the IXS to measure the dispersion and lifetime spectrum of an ScN film.

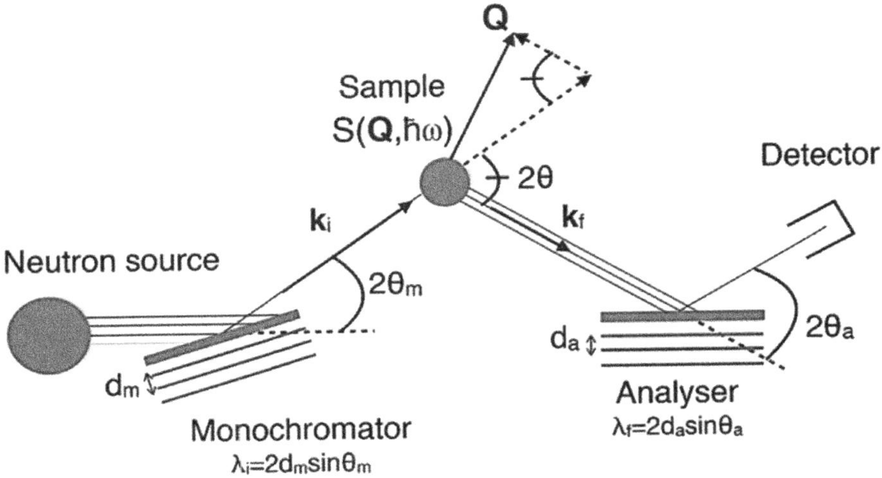

Figure 6.15. *Schematic layout of a neutron triple-axis spectrometer (TAS)*

6.2.4. *Phonon engineering for low thermal conductivity*

Because of the broad distribution of phonon MFPs and frequencies, effectively suppressing the overall heat conduction carried by various phonons is very challenging. Different methods based on multiple mechanisms should be applied simultaneously to achieve this goal (Volz et al. 2016; Nomura et al. 2018; Hori and Shiomi 2019). Qian et al. (2021, Figure 1) illustrated an overall picture of the underlying mechanism of phonon engineering for thermal conductivity control. For the low thermal conductivity required by thermoelectric materials, we can divide the approaches into three according to the relationship of phonon mean free path (MFP, Λ) and the characteristic length of the structure (d). For single-crystal or poly-crystal bulk material with large grain size ($d > \Lambda$), phonons transport diffusively and can be described by Fourier's law, hence named the Fourier diffusion regime. In the Fourier diffusion regime, lower thermal conductivity can be achieved by tailoring the phonon dispersion relation and phonon scattering rate by increasing the materials' anharmonicity or crystal complexity. If the characteristic length of nanostructures is smaller than MFP ($d < \Lambda$), phonons transport ballistically or quasi-ballistically and nanostructures such as interfaces, boundaries and precipitates significantly influence the thermal conductivity. This ballistic regime can be further divided into the Casimir–Knudsen and Coherent regimes based on whether the phonon phase is preserved or not when scattered by nanostructures.

In the Casimir–Knudsen regime, the low thermal conductivity is achieved by introducing extra phonon boundary scattering, leading to a shortening phonon relaxation time. There, the nanostructures do not alter the phonon states, i.e. dispersion relations and eigenmodes because the phase of the phonons is not preserved at the length scale of the nanostructures. This effect becomes predominant when the characteristic length becomes smaller, thus usually called the "size effect". If phonon scattering by the nanostructure is phase-preserving, phonons transport coherently and exhibit wave-like nature. In this coherent regime, the interaction between phonon and nanostructures can affect the dispersion relation of phonons, which can reduce the thermal conductivity by suppressing the phonon group velocity. In the coherent regime, we expect to see characteristics that reflect wave nature, such as interference, diffraction and localization. One of the extremes is the Anderson localization where the group velocities of various phonons become zero, exhibiting the characteristics of a standing wave. The challenge of coherent phonon engineering comes in because the coherent length of the phonon is extremely short and easily becomes decoherent. For example, a rough interface will easily destroy the wave nature of the phonon. Experimentally, due to the abovementioned issue, coherent phonon engineering is still challenging, where the effect is typically observed under low temperatures.

The following sections discuss some example of phonon engineering for improving thermoelectric materials.

6.2.5. *Intrinsic low thermal conductivity in complex lattice structure*

Reducing the intrinsic thermal conductivity of bulk materials, i.e. phonon engineering in the Fourier regime, can be achieved through reducing the group velocities and relaxation times by heavy element, weak bonding and complex unit cell structure. These approaches can be explained by Slack's early theory (Slack 1973) that describes the thermal conductivity of a material at a temperature point higher than the Debye temperature:

$$\kappa = A\frac{\overline{M}\sigma\theta^3}{\gamma T N^{2/3}} \qquad [6.32]$$

where A is the proportional constant, $\overline{M}$ is the mean atomic mass, σ^3 is the average volume of individual atoms, θ is the Debye temperature, γ is the Grüneisen parameter (γ is related to the phonon scattering), and N is the number of atoms in a unit cell. According to this equation, intrinsic low thermal conductivity can be achieved by heavy elements (large $\overline{M}$), or by weak interatomic potential, because θ is proportional to the group velocity $v = \sqrt{B\sigma/\overline{M}}$ (B is the bulk modulus). Representative examples are the commonly used commercial thermoelectric materials such as Bi_2Te_3 and PbTe, whose constituent elements are located after the fifth period of the elemental table, i.e. large atomic masses. Moreover, as the thermal conductivity is proportional to $N^{-2/3}$, a complex lattice structure with a large unit cell size such as $CoSb_3$ (Kawaharada et al. 2001) and Zn_4Sb_3 (Caillat et al. 1997) have low thermal conductivity.

However, Slack's theory still lacks details about the scattering term of phonons, which limits the control over the anharmonicity of phonons. This issue was solved with the development of the DFT calculations, which illustrate a clear picture of the harmonic phonon properties and provide fruitful information about the phonon scattering, especially for materials with a complex lattice structure, whose phonon properties are difficult to predict using the classic theory. With the help of DFT calculation, phonon anharmonicity such as the phonon rattling mode was found to lead to low thermal conductivity in cage-like structures, such as clathrate (Christensen et al. 2008) or skutterudite (Sales et al. 1996). The weak interaction between the intermediate heavy atoms and the host structure results in low-lying branches in phonon dispersion relations, which leads to scattering of acoustic branch phonons, and thus low thermal conductivity (Christensen et al. 2008, Figure 3(a)). Another example is the resonance mode in materials like PbTe, where the strong interaction between the acoustic and optical branches can lead to a negligible band gap between acoustic and optical phonons (Delaire et al. 2011). The strong tangling between the acoustic and optical phonons introduces significant anharmonicity, and thus low thermal conductivity.

Hybrid organic–inorganic crystals can also realize ultra-low thermal conductivity because of the significant difference in phonon properties and weak coupling between the organic network and the inorganic core molecules. For example, the hybrid structure of core nanoparticles (Au, CdSe, PbS, etc.) connected by ligand networks can give thermal conductivity as low as 0.01–0.03 W/m-K, comparable to air (Ong et al. 2013). Although these kinds of hybrid materials typically show low electrical conductance, which is currently unlikely to be applied as a thermoelectrical material, the situation may change because of the recent development of organic thermoelectric materials (Wang and Yu 2019).

6.2.6. *Low thermal conductivity by nanostructures*

According to Matthiessen's rule of phonon scattering, in addition to the Umklapp scattering that contributes to the intrinsic thermal conductivity of bulk materials, boundary scattering, mass-difference scattering and phonon–electron scattering also contribute to determining the thermal conduction. Note here, phonon–electron scattering is less important for typical thermoelectric materials because of low electrons density. Thermal conductivity suppression through boundary scattering and mass-difference scattering by nanostructures is the most commonly used method in the fabrication of thermoelectric materials.

The relaxation time of mass difference scattering rates is a function of phonon frequency ($\tau \sim \omega^4$), which indicates that high-frequency phonons are strongly scattered. Typically, this kind of mass difference can be introduced by the isotope mixture, alloying and impurities.

The interest in how the isotope mixture influences thermal conductivity arises when exploring both high and low thermal conductivity. For example, the thermal conductivity of c-BN is enhanced by 90% compared with nature c-BN, by alleviating the degree of the isotope mixture. Other isotope-mixture materials like GeC and BeSe also show significant enhancement in thermal conductivity when they are changed to be isotopically enriched (Lindsay et al. 2013). In the opposite direction, for existing materials that are isotropically enriched, such as graphite and carbon nanotubes (^{12}C enriched), introducing ^{13}C or ^{14}C can effectively decrease the thermal conductivity while maintaining the electrical conductance, which potentially enables the application as a thermoelectric material (Anno et al. 2014). Carbon nanotubes, whose Seebeck coefficient (S~170 μV/K) is comparable to Bi_2Te_3 (Nakai et al. 2014), can be potentially turned into a promising candidate for thermoelectrics when the original high thermal conductivity is reduced through the isotope mixture. In MD simulation studies, carbon nanotubes composed of a random mixture $^{13}C/^{12}C$ and periodic $^{13}C/^{12}C$ structure can effectively reduce the thermal conductivity by 30% and 56%, respectively, (Shiomi and Maruyama 2006). Graphene also shows enhanced thermoelectric performance by the isotope mixture, which causes the thermal conductivity to decrease to 10%, thus there is a 10 times enhancement in *ZT* (Anno et al. 2015).

Alloying is one of the most commonly used methods to reduce the thermal conductivity of thermoelectric materials. Similar to the isotope mixture method, it not only scatters high-frequency phonon by the mass disorder, but can also modify the dispersion of phonons. Murakami et al. (2013) found that the inhomogeneity of force fields introduced in $PbTe_{1-x}Se_x$ alloys significantly suppressed the thermal conductivity. In relation to this, Arrigoni et al. (2018) realized a similar effect of the inhomogeneity of force field introduced by the alloy method in $In_{1-x}Ga_xAs$ alloys. A representative case of the alloying method is the alloy between Sb_2Te_3, Bi_2Te_3 and Bi_2Se_3. Owing to a complete solid solubility of Bi_2Se_3, Sb_2Te_3 and Bi_2Te_3, they can form various alloys with high *ZT* by strongly suppressing the lattice thermal conductivity without degrading the electronic properties (Wright 1958; Kusano and Hori 2002). Many other thermoelectric materials like $PbTe/Sb_2Te_3$ (Zhu et al. 2005) and As/SiGe (Heller et al. 1976) alloys can also effectively suppress the thermal conductivity to achieve high *ZT*. In addition to the alloying method, which suppresses the thermal conduction of the high-frequency phonon by atomic-level mass disorder and inhomogeneity, disorder and nanostructures with relatively larger scale are necessary to further decrease the thermal conduction of low-frequency phonons. Theoretical studies indicate that nanoscale porosity can strongly scatter phonons to reduce the thermal conductivity, while showing a more negligible impact on electron scattering (Lee et al. 2010). The effect of nanoparticle scattering for lower thermal conductivity is also confirmed by theoretical calculation, similar to that of the nanoscale porosity (Zebarjadi et al. 2009).

Thermal conductivity suppression based on boundary scattering can be realized in a thin film, nanowire or composite with nanograins. Boundary scattering becomes predominant when the characteristic length of the nanostructure is smaller than the MFP of phonons. Boundary scattering is another widely used method to explore the low thermal for thermoelectric application. Promoted by the DFT calculation, which can precisely give the MFP of both phonon and electrons, the strategy is to fabricate nanomaterials with a characteristic length longer than the MFP of phonon and shorter than that of the electron. The thermal conductivity of silicon can be reduced by approximately 100 times due to the significant boundary scattering in nanowires (Boukai et al. 2008; Hochbaum et al. 2008). In composite materials with a nano-size grain boundary, interfacial phonon scattering and transmission becomes important. Poudel et al. (2008) realized high *ZT* of nanostructured BiSbTe caused by the phonon scattering across the nanograin boundary. Miura et al. (2015) suppressed the thermal conductivity of silicon by sintering nanoparticles, where small grain size and the resultant high interface density significantly reduced the thermal conductivity and enhanced *ZT*. As the interfacial thermal conductance varies significantly with the interfacial conditions, controlling the interfacial structure by introducing disorder or stress at the interface effectively introduces thermal resistance in the composite. The effectiveness of interfacial thermal resistance in the total thermal conductivity can be studied using the effective medium theory (EMF). Here is a mostly used model of composite materials proposed by Nan et al. (1997):

$$\kappa_{\mathrm{eff}} = \kappa_{\mathrm{m}} \frac{\kappa_{\mathrm{p}}(1+2\alpha)+2\kappa_{\mathrm{m}}+2f\left[\kappa_{\mathrm{p}}(1-\alpha)-\kappa_{\mathrm{m}}\right]}{\kappa_{\mathrm{p}}(1+2\alpha)+2\kappa_{\mathrm{m}}-f\left[\kappa_{\mathrm{p}}(1-\alpha)-\kappa_{\mathrm{m}}\right]} \qquad [6.33]$$

$$\alpha = \frac{\kappa_{\mathrm{m}}}{G\cdot r} \qquad [6.34]$$

where subscripts eff, m, and p denote the composites, matrix and filler, respectively, f is the volume fraction of the filler, r is the radius of spherical filler and G is the interfacial thermal resistance. According to this equation, the thermal boundary resistance becomes critical when the grain size is nano-order. For example, in the composite materials fabricated by sintering silicon nanowires, the thermal conductivity decreases significantly due to the interfacial thermal conductance reduction caused by the interfacial softening and gives a high *ZT* of 0.3 at room temperature (Kashiwagi et al. 2019). To better understand how the interface influences the overall thermal conductivity of thermoelectric silicon composite, Sakata et al. (2015) directly measured the interfacial thermal conductance between silicons fabricated by the sintering bonding method. The bonded silicon interface can mimic the nano-interfaces of silicon composite well (Figure 6.16), which indicates the oxidation degree can enormously reduce the interfacial thermal conductance.

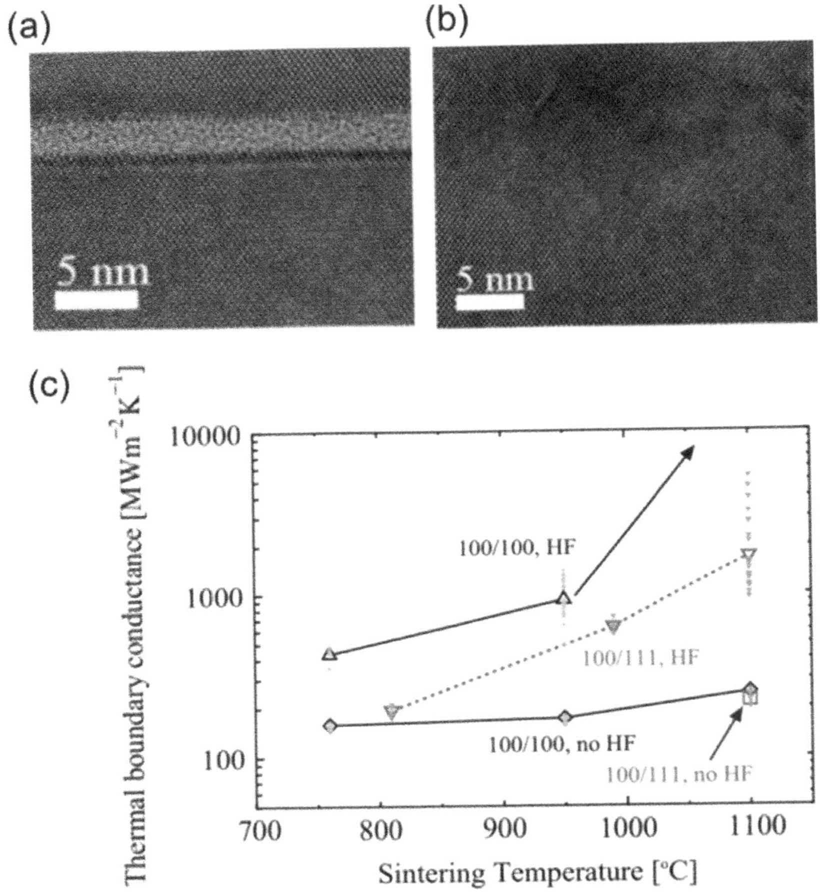

Figure 6.16. *(a) and (b) TEM images of the sintering bonded silicon interface with high and low oxidation degree. (c) Thermal boundary conductance variation as a function of the oxidation and sintering parameters. Figure adapted with permission from Sakata et al. (2015). For a color version of this figure, see www.iste.co.uk/akinaga/thermoelectric2.zip*

6.2.7. *Coherent phonon engineering in superlattice*

In contrast with the inelastic scattering in which phonon interaction with nanoparticles or interfaces is phase-destroying, coherent phonon transport is ballistic and passes through the interface in a preserved phase. Since coherent phonon transport can be regarded as a propagating wave, phonon interaction with the interface/nanoparticles can be viewed as wave diffraction and interference (Maldovan 2015). In analogy to photons and electrons, phonons exhibit a wave-like nature when the characteristic length of the structure is comparable with the wavelength (Figure 6.17(a)). Take light as an example, when the layer thickness is similar to the $\lambda/2$ wavelength, light intensity is suppressed due to wave interference. In this sense, phonon interference can create forbidden energy bands in phonon dispersion and strongly suppress the thermal conductivity of a particular phonon (Figure 6.17(b)).

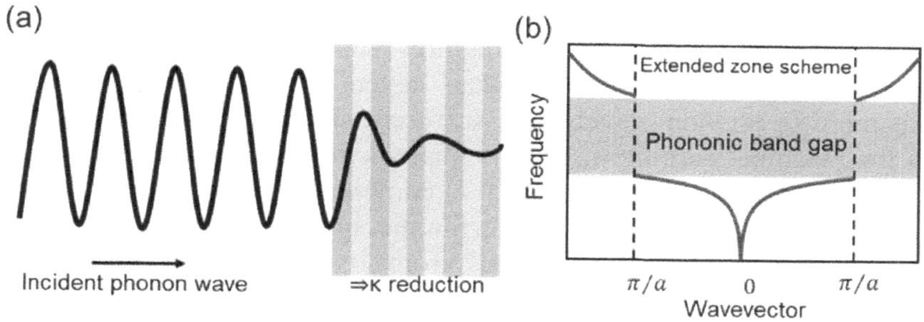

Figure 6.17. *(a) Schematic of phonon interference, and the resultant forbidden energy bands in the phonon dispersion relation. For a color version of this figure, see www.iste.co.uk/akinaga/thermoelectric2.zip*

Early works on coherent phonon engineering are usually realized by layered superlattice or other periodic structures. Note here, multi-interfaces are typically used to realize substantial phonon interference through multiple transmission/reflection at the interface. Phonon interference can result in flat bands in phonon dispersion relation, which means low group velocity. The phonon engineering of the coherent phonon is similar to the band engineering for bulk material, which modifies phonon harmonic properties. There have been many computational works studying coherent phonon engineering. For example, the multilayer structure of germanium nanoparticles embedded silicon matrix was found to cause collective phonon resonances in the embedded germanium nanoparticles region (Feng et al. 2017). Experimentally, as directly measuring the wave-behavior of individual phonons or the dispersion relation are technically challenging, coherent phonon behavior is typically confirmed by temperature dependence thermal conductivity measurement of the structure with various periods and individual layer thickness. Ravichandran et al. (2014) observed the coherent phonon regime in $CaTiO_2/SrTiO_2$ superlattice when the thickness of individual periods is smaller than 1 nm. Maire et al. (2017) modified the in-plane thermal conductivity of Si thin film fabricated with nanosize holes through tuning the coherent phonon transport. Luckyanova et al. (2012) observed coherent phonon behavior in GaAs/AlAs superlattice under 150 K, and a transition of phonons from coherent to diffusive in GaAs/AlAs superlattices with randomly distributed ErAs nanodots in a wide temperature range from 30–300 K (Luckyanova et al. 2018).

On the contrary, simultaneously reducing the thermal conductivity of phonons of various wavelengths requires superlattice structures with an aperiodic structure whose individual layers have various thicknesses comparable to phonon wavelength. In recent studies, simulation combined with the machine learning method is implemented to determine the optimal aperiodic superlattices. Ju et al. (2017) combined the Green's function and the Bayesian optimization to search for the structure of Ge/Si superlattice with minimal thermal conductivity from over 60,000 candidates, and realized a 50% reduction of thermal conductivity compared with the periodic structure by structure optimization. Experimental demonstration of the ultra-low thermal conductivity in aperiodic GaAs/AlAs superlattice optimized through machine-learning (Figure 6.18) has also realized a large reduction in thermal conductivity than that of the periodic superlattice. The simulation shows good techniques with the experimental result and indicates that the origin of the decrease in thermal conductivity comes from the phonon localization (Hu et al. 2020). Notably, the coherent phonon behavior is typically observed under a low-temperature regime because low temperature reduces the possibility of inelastic scattering of phonons. The challenge comes in realizing the coherent phonon control under room temperature or higher temperature, which is essential for thermoelectric application. Because coherent phonon engineering requires an ultra-fine structure comparable to the phonon wavelength, which is typically atomic-scale, the resultant large interface density causes a higher possibility of inelastic scattering owing to inevitably introduced defects or interfacial roughness during the fabrication. Besides, the superlattice materials are limited to those with similar lattice structures, as a large lattice mismatch at the interface would result in defects and disorders and destroy the phonon technique.

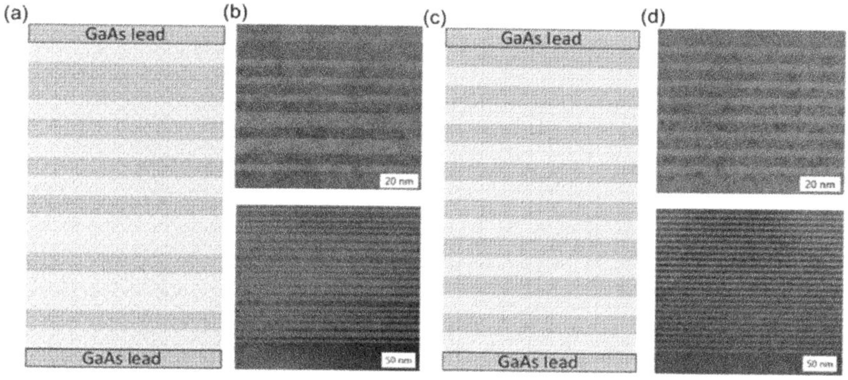

Figure 6.18. *The structure and TEM image of GaAs/AlAs (a) and (b) periodic structure and (c) and (d) aperiodic structure optimized through machine learning. Figure adapted with permission from Hu et al. (2020). For a color version of this figure, see www.iste.co.uk/akinaga/thermoelectric2.zip*

Recently, the van der Waals stacking structure by two-dimensional materials emerged as a perfect candidate to realize the phonon engineering technique. The atomic-level flatness/thickness can maintain phonon coherence. The van der Waals inter-layer interaction enables assembling materials with distinctive lattice structures, resulting in a strong interference effect. The work by Hu et al. (2021), based on MD simulation and machine learning, has indicated the strong localization for various phonons in a graphene/MoS_2 aperiodic superlattice structure (Figure 6.19). Experimentally, ultra-low cross-plane thermal conductivity is confirmed in structures such as the Graphene/$MoSe_2$/MoS_2/Wse_2 stacking structure (Vaziri et al. 2019) and randomly stacking MoS_2 superlattice (Kim et al. 2021). However, the underlying mechanism is still under debate, whether the low thermal conductivity originates from the coherent phonon interference or the interfacial inelastic phonon scattering. Another challenge lies in fabricating multilayer structures with large scale, i.e. large numbers of layers and large areas, and elucidating the correlation between structure and coherent phonon control, i.e. how the localization is enhanced by structure optimization.

Overall, coherent phonon engineering is an ideal approach for thermoelectrics, by realizing low thermal conductivity without decreasing the electrical conductance. At the same time, severe challenges still require further development both theoretically and methodologically.

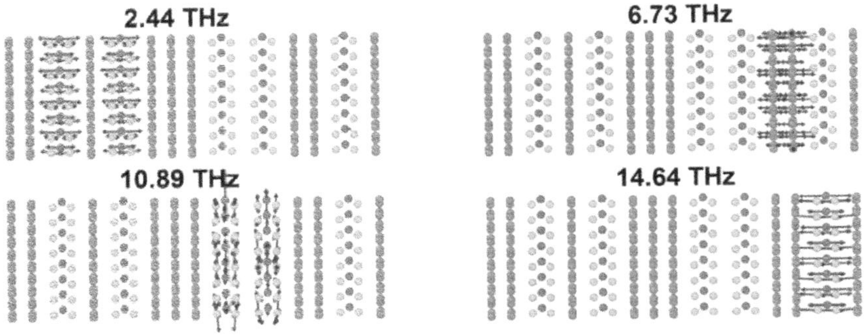

Figure 6.19. *Localized phonon modes in the aperiodic MoS2/graphene superlattice. Figure adapted with permission from Hu et al. (2021). For a color version of this figure, see www.iste.co.uk/akinaga/thermoelectric2.zip*

6.3. Summary and future prospects

In this chapter, keeping in mind the need for low thermal conductivity in thermoelectrics, following the details about the fundamentals of thermal conduction and phonons, we comprehensively introduced methods of thermal conductivity measurement and phonon properties analysis and the basic principles and application examples of phonon engineering.

From the methodological aspects of phonon properties and thermal conductivity analysis, experimental techniques allow us to measure the thermal conductivity from nano- to macro-scales accurately. In addition, further development based on nanofabrication, frequency modulation and dual laser interference techniques provides an indirect approach to unravel the anharmonic properties of phonons. Temperature-dependent thermal conductivity measurements indirectly reflect the contribution of different phonon transport mechanisms to thermal conductivity in a given material. On a related point, both harmonic and anharmonic properties of phonons can be directly probed by X-ray, neutron, photon and electron diffraction methods. On the one hand, computational methods based on first principles, lattice dynamics and molecular dynamics, etc., allow us not only to predict the thermal conductivity of materials or structures at various scales, but also to obtain massive information about phonon transport, such as states, velocities, relaxation times, mean free paths, all in a mode-dependent fashion, which are essential in explaining the experimental result and providing a guideline for effective optimization on the lattice or nanostructures for low thermal conductivity.

As for the mechanisms and strategies of phonon engineering, low thermal conductivity can be realized both intrinsically and extrinsically by various approaches. The complex lattice structure, large unit cell, weak interatomic interaction and large atomic mass can result in low phonon group velocity and significant Umklapp scattering, strongly suppressing the intrinsic thermal conductivity. Alloying or isotropic mixtures and nanostructuring with grains, pores and precipitates can further enhance extrinsic phonon scattering or manipulate the phonon wave-behavior in case phonon phases are preserved to reduce thermal conductivity.

While the remarkable progress in phonon engineering has promoted the development of thermoelectric materials, there is still plenty of room for advancement. Some examples for potential advancement are as follows.

The interfacial state could be further incorporated into phonon engineering. In the Casimir–Knudsen regime, phonon transport is usually characterized by dividing a material into the internal region (e.g. inside a grain) and sharp interface (e.g. grain boundary). However, the influence of interface to phonon states is known to extend inside the internal region. This extension length strongly depends on the phonon mode and can be non-negligible compared with the nanostructure length scale (e.g. grain size) for low frequency and long wavelength acoustic phonons. In addition, in the actual materials, the interface usually takes a complex structure and composition, which makes this issue even more complicated. While this calls for more advancement in the theory of phonon transport, the presence of multiple length scales can also be seen as an opportunity to gain controllability. Like the well-known concept of improving thermoelectric ZT by nanostructuring using the difference in mean free paths of electrons and phonons, i.e. to scatter phonons dominantly by setting the nanostructure length scale in between the electron and phonon mean free paths, the added degrees of freedom by the interfacial states may be used to separately control thermal conductivity, electrical conductivity and the Seebeck coefficient.

Better usage of phonon coherence by nanostructures is also an issue that has been discussed for a long time without being resolved. It is now clear that the requirements to do this at room temperature is severe because phonon coherence is quite vulnerable against interface roughness due to the short nominal wavelength. When material fabrication technology greatly advances and enables the fabrication of bulk scale nanostructured materials with atomically smooth interfaces, coherent phonon engineering is expected to become extremely powerful. An optimistic view of this is that this has already been realized is some cases; large unit cell compounds (e.g. manganese silicide) are in a way self-assembled nanostructures with atomically smooth interfaces, and more nanostructures of such may be produced by looking into the metastable phase further. There, the currently growing materials informatics techniques should be useful to look for new metastable structures.

In the diffusion regime, the studies on highly anharmonic materials so far have identified some key quantum features that give rise to strong anharmonicity like electrical polarization, electron lone pairs and resonance bonding. The next step would be to design and actually fabricate materials with the features. Again, expanding the search space to the metastable phase should help. It should be noted that when applying thermoelectrics for powering IoT sensors/transmitters, the temperature of interest is around room temperature, which makes the thermal stability issues of metastable materials less important.

Clearly, all of the above require a research platform that closely couples fabrication, solid-state theory, machine learning, numerical simulation and measurements, and thus calls for extensive interdisciplinary and international collaborations among researchers.

6.4. References

Anno, Y., Takei, K., Akita, S., Arie, T. (2014). Artificially controlled synthesis of graphene intramolecular heterojunctions for phonon engineering. *Physica Status Solidi – Rapid Research Letters*, 8(8), 692–697.

Anno, Y., Takei, K., Akita, S., Arie, T. (2015). Enhancing the thermoelectric device performance of graphene using isotopes and isotopic heterojunctions. *Advanced Electronic Materials*, 1(9), 1–6.

Arrigoni, M., Carrete, J., Mingo, N., Madsen, G.K.H. (2018). First-principles quantitative prediction of the lattice thermal conductivity in random semiconductor alloys: The role of force-constant disorder. *Physical Review B*, 98(11), 115205.

Ashcroft, N.W. and Mermin, N.D. (1976). *Solid State Physics*. Holt-Saunders, Philadelphia.

Attig, N., Binder, K., Grubmueller, H., Kremer, K. (2004). Publication series of the John von Neumann Institute for Computing (NIC). *NIC Series*, 22.

Basko, D.M., Piscanec, S., Ferrari, A.C. (2009). Electron-electron interactions and doping dependence of the two-phonon Raman intensity in graphene. *Physical Review B*, 80(16), 165413.

Boukai, A.I., Bunimovich, Y., Tahir-Kheli, J., Yu, J.K., Goddard, W.A., Heath, J.R. (2008). Silicon nanowires as efficient thermoelectric materials. *Nature*, 451(7175), 168–171.

Broido, D.A., Malorny, M., Birner, G., Mingo, N., Stewart, D.A. (2007). Intrinsic lattice thermal conductivity of semiconductors from first principles. *Applied Physics Letters*, 91(23), 19–22.

Cahill, D.G. (1990). Thermal conductivity measurement from 30 to 750 K: The 3ω method. *Review of Scientific Instruments*, 61(2), 802–808.

Cahill, D.G. (2004). Analysis of heat flow in layered structures for time-domain thermoreflectance. *Review of Scientific Instruments*, 75(12), 5119–5122.

Caillat, T., Fleurial, J.P., Borshchevsky, A. (1997). Preparation and thermoelectric properties of semiconducting Zn_4Sb_3. *Journal of Physics and Chemistry of Solids*, 58(7), 1119–1125.

Christensen, M., Abrahamsen, A.B., Christensen, N.B., Juranyi, F., Andersen, N.H., Lefmann, K., Andreasson, J., Bahl, C.R.H., Iversen, B.B. (2008). Avoided crossing of rattler modes in thermoelectric materials. *Nature Materials*, 7(10), 811–815.

Delaire, O., Ma, J., Marty, K., May, A.F., McGuire, M.A., Du, M.H., Singh, D.J., Podlesnyak, A., Ehlers, G., Lumsden, M.D. et al. (2011). Giant anharmonic phonon scattering in PbTe. *Nature Materials*, 10(8), 614–619.

Esfarjani, K., Chen, G., Stokes, H.T. (2011). Heat transport in silicon from first-principles calculations. *Physical Review B*, 84(8), 085204.

Feng, L., Shiga, T., Han, H., Ju, S., Kosevich, Y.A., Shiomi, J. (2017). Phonon-interference resonance effects by nanoparticles embedded in a matrix. *Physical Review B*, 96(22), 1–5.

Ferraro, J.R., Nakamoto, K., Brown, C.W. (1994). *Introductory Raman Spectroscopy*. Elsevier, Amsterdam.

Gubaev, K., Podryabinkin, E.V., Hart, G.L.W., Shapeev, A.V. (2019). Accelerating high-throughput searches for new alloys with active learning of interatomic potentials. *Computational Materials Science*, 156(September 2018), 148–156.

Heller, M.W., Nasby, R.D., Johnson, R.T. (1976). Electrical transport properties of SiGe thermoelectric alloys doped with As, P, and As+P. *Journal of Applied Physics*, 47(9), 4113–4121.

Hochbaum, A.I., Chen, R., Delgado, R.D., Liang, W., Garnett, E.C., Najarian, M., Majumdar, A., Yang, P. (2008). Enhanced thermoelectric performance of rough silicon nanowires. *Nature*, 451(7175), 163–167.

Hori, T. and Shiomi, J. (2019). Tuning phonon transport spectrum for better thermoelectric materials. *Science and Technology of Advanced Materials*, 20(1), 10–25.

Hu, R., Iwamoto, S., Feng, L., Ju, S., Hu, S., Ohnishi, M., Nagai, N., Hirakawa, K., Shiomi, J. (2020). Machine-learning-optimized aperiodic superlattice minimizes coherent phonon heat conduction. *Physical Review X*, 10(2), 21050.

Hu, S., Ju, S., Shao, C., Guo, J., Xu, B., Ohnishi, M., Shiomi, J. (2021). Ultimate impedance of coherent heat conduction in van der Waals graphene-MoS2 heterostructures. *Materials Today Physics*, 16, 100324.

Jiang, P., Qian, X., Yang, R. (2018). Tutorial: Time-domain thermoreflectance (TDTR) for thermal property characterization of bulk and thin film materials. *J. Appl. Phys.*, 124(16), 161103.

Ju, S., Shiga, T., Feng, L., Hou, Z., Tsuda, K., Shiomi, J. (2017). Designing nanostructures for phonon transport via Bayesian optimization. *Physical Review X*, 7(2), 021024.

Kashiwagi, M., Liao, Y., Ju, S., Miura, A., Konishi, S., Shiga, T., Kodama, T., Shiomi, J. (2019). Scalable multi-nanostructured silicon for roomerature thermoelectrics. *ACS Applied Energy Materials*, 2(10), 7083–7091.

Kawaharada, Y., Kurosaki, K., Uno, M., Yamanaka, S. (2001). Thermoelectric properties of $CoSb_3$. *Journal of Alloys and Compounds*, 315(1–2), 193–197.

Kim, S.E., Mujid, F., Rai, A., Eriksson, F., Suh, J., Poddar, P., Ray, A., Park, C., Fransson, E., Zhong, Y. et al. (2021). Extremely anisotropic van der Waals thermal conductors. *Nature*, 597(7878), 660–665.

Kittel, C., McEuen, P., McEuen, P. (1996). *Introduction to Solid State Physics*. Wiley, New York.

Klemens, P.G. (1955). The scattering of low-frequency lattice waves by static imperfections. *Proceedings of the Physical Society. Section A*, 68(12), 1113–1128.

Kohn, W. (1959). Image of the fermi surface in the vibration spectrum of a metal. *Physical Review Letters*, 2(9), 393–394.

Kulda, J., Strauch, D., Pavone, P., Ishii, Y. (1994). Inelastic-neutron-scattering study of phonon eigenvectors and frequencies in Si. *Physical Review B*, 50(18), 13347–13354.

Kusano, D. and Hori, Y. (2002). Thermoelectric properties of p-type $(Bi_2Te_3)_{0.2}(Sb_2Te_3)_{0.8}$ thermoelectric material doped with PbTe. *Nippon Kinzoku Gakkaishi/Journal of the Japan Institute of Metals*, 66(10), 1063–1065.

Lee, H., Vashaee, D., Wang, D.Z., Dresselhaus, M.S., Ren, Z.F., Chen, G. (2010). Effects of nanoscale porosity on thermoelectric properties of SiGe. *Journal of Applied Physics*, 107(9).

Lindsay, L., Broido, D.A., Reinecke, T.L. (2013). Phonon-isotope scattering and thermal conductivity in materials with a large isotope effect: A first-principles study. *Physical Review B – Condensed Matter and Materials Physics*, 88(14), 1–9.

Luckyanova, M.N., Garg, J., Esfarjani, K., Jandl, A., Bulsara, M.T., Schmidt, A.J., Minnich, A.J., Chen, S., Dresselhaus, M.S., Ren, Z. et al. (2012). Coherent phonon heat conduction in superlattices. *Science*, 338(6109), 936–939.

Luckyanova, M.N., Mendoza, J., Lu, H., Song, B., Huang, S., Zhou, J., Li, M., Dong, Y., Zhou, H., Garlow, J. et al. (2018). Phonon localization in heat conduction. *Science Advances*, 4(12), eaat9460.

Maire, J., Anufriev, R., Yanagisawa, R., Ramiere, A., Volz, S., Nomura, M. (2017). Heat conduction tuning by wave nature of phonons. *Science Advances*, 3(8), 1–7.

Maldovan, M. (2015). Phonon wave interference and thermal bandgap materials. *Nature Materials*, 14(7), 667–674.

McGaughey, A.J. and Larkin, J.M. (2014). Predicting phonon properties from equilibrium molecular dynamics simulations. *Annual Review of Heat Transfer*, 17.

Miura, A., Zhou, S., Nozaki, T., Shiomi, J. (2015). Crystalline-amorphous silicon nanocomposites with reduced thermal conductivity for bulk thermoelectrics. *ACS Applied Materials and Interfaces*, 7(24), 13484–13489.

Moon, I.K., Jeong, Y.H., Kwun, S.I. (1996). The 3ω technique for measuring dynamic specific heat and thermal conductivity of a liquid or solid. *Review of Scientific Instruments*, 67(1), 29–35.

Mortazavi, B., Podryabinkin, E.V., Roche, S., Rabczuk, T., Zhuang, X., Shapeev, A.V. (2020). Machine-learning interatomic potentials enable first-principles multiscale modeling of lattice thermal conductivity in graphene/borophene heterostructures. *Materials Horizons*, 7(9), 2359–2367.

Murakami, T., Shiga, T., Hori, T., Esfarjani, K., Shiomi, J. (2013). Importance of local force fields on lattice thermal conductivity reduction in PbTe1−xSex alloys. *Europhysics Letters*, 102(4), 46002.

Nakai, Y., Honda, K., Yanagi, K., Kataura, H., Kato, T., Yamamoto, T., Maniwa, Y. (2014). Giant Seebeck coefficient in semiconducting single-wall carbon nanotube film. *Applied Physics Express*, 7(2), 025103.

Nan, C.W., Birringer, R., Clarke, D.R., Gleiter, H. (1997). Effective thermal conductivity of particulate composites with interfacial thermal resistance. *Journal of Applied Physics*, 81(10), 6692–6699.

Nomura, M., Shiomi, J., Shiga, T., Anufriev, R. (2018). Thermal phonon engineering by tailored nanostructures. *Japanese Journal of Applied Physics*, 57(8).

Ong, W.L., Rupich, S.M., Talapin, D.V., McGaughey, A.J.H., Malen, J.A. (2013). Surface chemistry mediates thermal transport in three-dimensional nanocrystal arrays. *Nature Materials*, 12(5), 410–415.

Peterson, R.E. and Anderson, A.C. (1972). Acoustic-mismatch model of the Kaptiza resistance. *Physics Letters A*, 40(4), 317–319.

Poudel, B., Hao, Q., Ma, Y., Lan, Y., Minnich, A., Yu, B., Yan, X., Wang, D., Muto, A., Vashaee, D. et al. (2008). High-thermoelectric performance of nanostructured bismuth antimony telluride bulk alloys. *Science*, 320(5876), 634–638.

Qian, X., Zhou, J., Chen, G. (2021). Phonon-engineered extreme thermal conductivity materials. *Nature Materials*, 20(9), 1188–1202.

Ravichandran, J., Yadav, A.K., Cheaito, R., Rossen, P.B., Soukiassian, A., Suresha, S.J., Duda, J.C., Foley, B.M., Lee, C.H., Zhu, Y. et al. (2014). Crossover from incoherent to coherent phonon scattering in epitaxial oxide superlattices. *Nature Materials*, 13(2), 168–172.

Sakata, M., Hori, T., Oyake, T., Maire, J., Nomura, M., Shiomi, J. (2015). Tuning thermal conductance across sintered silicon interface by local nanostructures. *Nano Energy*, 13, 601–608.

Sales, B.C., Mandrus, D., Williams, R.K. (1996). Filled skutterudite antimonides: A new class of thermoelectric materials. *Science*, 272(5266), 1325–1328.

Shiomi, J. (2014). Nonequilirium molecular dynamics methods for lattice heat conduction calculations. *Annual Review of Heat Transfer*, 17.

Shiomi, J. and Maruyama, S. (2006). Heat conduction of single-walled carbon nanotube isotope superlattice structures: A molecular dynamics study. *Physical Review B – Condensed Matter and Materials Physics*, 74(15), 1–6.

Slack, G.A. (1973). Nonmetallic crystals with high thermal conductivity. *Journal of Physics and Chemistry of Solids*, 34(2), 321–335.

Swartz, E.T. and Pohl, R.O. (1989). Thermal boundary resistance. *Reviews of Modern Physics*, 61(3), 605–668.

Tian, Z., Garg, J., Esfarjani, K., Shiga, T., Shiomi, J., Chen, G. (2012). Phonon conduction in PbSe, PbTe, and $PbTe_{1-x}Se_x$ from first-principles calculations. *Physical Review B – Condensed Matter and Materials Physics*, 85(18), 184303.

Uchiyama, H., Oshima, Y., Patterson, R., Iwamoto, S., Shiomi, J., Shimamura, K. (2018). Phonon lifetime observation in epitaxial ScN film with inelastic X-Ray scattering spectroscopy. *Physical Review Letters*, 120(23), 235901.

Vaziri, S., Yalon, E., Rojo, M.M., Suryavanshi, S.V., Zhang, H., McClellan, C.J., Bailey, C.S., Smithe, K.K.H., Gabourie, A.J., Chen, V. et al. (2019). Ultrahigh thermal isolation across heterogeneously layered two-dimensional materials. *Science Advances*, 5(8), 1–8.

Volz, S., Shiomi, J., Nomura, M., Miyazaki, K. (2016). Heat conduction in nanostructured materials. *Journal of Thermal Science and Technology*, 11(1), JTST0001–JTST0001.

Wang, H. and Yu, C. (2019). Organic thermoelectrics: Materials preparation, performance optimization, and device integration. *Joule*, 3(1), 53–80.

Wright, D.A. (1958). Thermoelectric properties of bismuth telluride and its alloys. *Nature*, 181, 834.

Yan, X., Liu, W., Wang, H., Chen, S., Shiomi, J., Esfarjani, K., Wang, H., Wang, D., Chen, G., Ren, Z. (2012). Stronger phonon scattering by larger differences in atomic mass and size in p-type half-Heuslers $Hf_{1-x}Ti_xCoSb_{0.8}Sn_{0.2}$. *Energy & Environmental Science*, 5(6), 7543.

Yang, L., Huh, D., Ning, R., Rapp, V., Zeng, Y., Liu, Y., Ju, S., Tao, Y., Jiang, Y., Beak, J. et al. (2021). High thermoelectric figure of merit of porous Si nanowires from 300 to 700 K. *Nature Communications*, 12(1), 3926.

Zebarjadi, M., Esfarjani, K., Shakouri, A., Bahk, J.H., Bian, Z., Zeng, G., Bowers, J., Lu, H., Zide, J., Gossard, A. (2009). Effect of nanoparticle scattering on thermoelectric power factor. *Applied Physics Letters*, 94(20), 202105.

Zhang, L., Han, J., Wang, H., Car, R., Weinan, E. (2018). Deep potential molecular dynamics: A scalable model with the accuracy of quantum mechanics. *Physical Review Letters*, 120(14), 143001.

Zhao, L.-D., Lo, S.-H., Zhang, Y., Sun, H., Tan, G., Uher, C., Wolverton, C., Dravid, V.P., Kanatzidis, M.G. (2014). Ultralow thermal conductivity and high thermoelectric figure of merit in SnSe crystals. *Nature*, 508(7496), 373–377.

Zhu, P., Imai, Y., Isoda, Y., Shinohara, Y., Jia, X., Zou, G. (2005). Enhanced thermoelectric properties of PbTe alloyed with Sb_2Te_3. *Journal of Physics Condensed Matter*, 17(46), 7319–7326.

PART 2

Toward Device Applications

7

The Current State of Thermoelectric Technologies and Applications with Prospects

Slavko BERNIK
Jožef Stefan Institute, Department for Nanostructured Materials, Ljubljana, Slovenia

7.1. Introduction

Water, food and energy are the basics necessary for survival. If you think primarily of water and food, when it comes to survival, then you likely live in a place on planet Earth, where temperatures are high enough all year round so that you do not need to heat your living space (unlike most of the human population). In any case, energy in all its forms is one of the foundations of prosperity and the key condition to be able to use and exploit on a daily basis the advantages of modern technology, which contributes to our quality of life. Good (i.e. healthy) food and pure drinking water are the qualities required of these substances, but what about quality of our energy? Certainly, it is important for energy to be available when needed, in the right form, in sufficient amounts, and equally importantly, for an acceptable price.

Thermoelectric Micro/Nano Generators 2,
coordinated by Hiroyuki AKINAGA, Atsuko KOSUGA,
Takao MORI and Gustavo ARDILA.

And this must continue to be applied in the future. After all, the affordability (i.e. availability, price) of energy strongly affects socio-economic welfare and the poverty threshold. However, with the paradigm of sustainability, another quality factor, related to the way that energy is produced, and the impact on the environment is becoming increasingly important. The requirements for energy will continue to grow, despite the global efforts for green development and the emphasis on effective and wise consumption. This will be related, to some extent, to the growth of the human population as well as an increasing energy consumption per person. Accordingly, the problem of energy delivery will continue to grow. Hence, the energy challenges of our civilization will not be solved without the use of diversified, renewable energy sources and enabling technologies, which increase current energy production and, to the greatest extent, replace fossil fuels and other conventional sources. Importantly, we have the knowledge and technologies to enable renewable energy in the forms of light, wind, heat, motion, etc., which can be converted into electricity. Electrical energy will certainly be the most desired form of energy in the future and the type of energy with the fastest demand growth. Hence, it will account for the largest share of the growing needs for energy. Solar and wind are the most popular and so far, only renewable energy sources exploited for the large-scale harvesting of electrical energy using the technologies of photovoltaic panels and wind-turbine generators, respectively (Ellabban et al. 2014; Abas et al. 2015; Ahmad and Zhang 2020; Paravantis and Kontouli 2020; Salvardi and Salvardi 2020; Stančin et al. 2020).

Heat is another renewable source of energy that provides great possibilities and should be exploited in the best possible way. Considering that the heat dissipated into the environment during the primary production of energy from non-renewable fossil fuels accounts for almost two-thirds of the produced energy, the conversion of just a small fraction of the wasted heat into electricity could contribute a significant amount of energy. Furthermore, we could use any source of heat in our environment to extract electrical energy, which brings tremendous possibilities. After all, the amount of energy that is required for a particular application or function can vary by 10 or more orders of magnitude in the range from micro-W (μW) to mega-W (MW). And such needs for energy are precisely in line with the nature of the heat sources, being widely dispersed and greatly varying in terms of the power of the source; with the exception of geothermal heat, high-, medium- and low-power heat sources are mostly related with the human activities (power plants, manufacturing facilities, cars, households, etc.), while natural low- to very-low-power heat sources are essentially everywhere in our environment

(Kishore and Priya 2018; Zabek and Morini 2019). Actually, we humans are a source of energy, emitting about 100 W of heat while resting and 525 W during physical effort. This can also be used to generate electricity. In fact, there are numerous examples of devices powered by human-body heat using thermoelectric technologies which exploit the Seebeck effect to convert heat into electricity (Siddique et al. 2017; Nozariasbmarz et al. 2019; Nozariasbmarz et al. 2020a). The large-scale energy harvesting of waste heat sources could definitely make a significant contribution to covering our growing energy needs. Micro-energy harvesting can probably be neglected in the big picture; however, its benefits, which have been increasingly recognized in recent years, lie elsewhere. With the rapid development of low-power applications in the fields of sensors, MEMS, biomedical devices, etc., and in the concept of the Internet of Things (IoT), the possibility to extract energy from the environment and to make all these devices energy autonomous and wirelessly connected to a remote post, even without batteries (which have to be replaced) become extremely interesting. Initial predictions for the entry into era of a sophisticated society with sensing of the environment using a network consisting of a trillion sensors via IoT technology were quickly faced with the impossible reality of how to connect each sensor individually to a power source. The other option is batteries, which are convenient (i.e. available) despite their disadvantages in terms of power, cost and maintenance (i.e. replacement). However, on the predicted scale of a trillion sensors in the network, it is practically impossible to keep replacing the batteries connected to the sensors. To realize this, the vision of a sophisticated societal implementation of an energy-harvesting technology is becoming vital. Such micro-energy harvesting can include solar, thermal, kinetic, electromagnetic and other environmental energy sources, exploiting the photovoltaic, electromagnetic, piezoelectric, triboelectric, electrostatic, electrostrictive, pyroelectric, magnetostrictive, dielectric and thermoelectric energy-conversion principles. In recent years, we have been witnessing tremendous progress in the development of energy-harvesting devices, including individual single-source devices and devices with hybrid configurations for single and multiple energy sources combining different conversion principles (Bai et al. 2018; Prauzek et al. 2018; Ryu et al. 2019; Akinaga 2020; Charris et al. 2020; Liu et al. 2021). These hybrid harvesting devices are certainly more complex than single-source devices; however, their advantage concerns the ability to exploit various energy sources simultaneously or each of them when available, thus increasing the reliability of the power supply around the clock. A nice example is a thermoelectric–photovoltaic hybrid system in which the efficiency of a wide-band-gap solar cell can be enhanced by up to 3.1% using a thermoelectric generator recovering the heat accumulated in the solar cell to produce additional power (Lorenzi et al. 2021). In the case of a single energy source, hybrid harvesters use different energy-conversion methods for better efficiency. In general, it is reasonable to use any source of energy available, even better if it is steady and reliable. Such heat sources are mostly available everywhere, and hence heat conversion is often one of the elements of hybrid harvesting devices. In any case, the basis is the development and technological progress of single-source harvesting devices with enhanced conversion efficiencies, which would also enable further progress with hybrid devices.

Thermoelectric technologies enable the direct conversion of heat into electricity via the Seebeck effect, regardless of the large dispersion of heat sources and the broad variations in their power outputs, while by applying an electric current via the Peltier effect, the heat can be transferred for either cooling or heating, depending on the direction of the current. It should be emphasized that no other technology is capable of this. Furthermore, both functions, i.e. energy harvesting and cooling, are without any moving parts (solid-state construction and operation) and without the production of environmentally hazardous waste, while maintenance is also not needed. So, why are thermoelectric technologies basically unknown to the general public, while almost everyone knows about photovoltaic solar panels and wind turbines? This issue is all the more relevant given the many success stories using thermoelectric technologies for both energy harvesting and cooling. Actually, a quick look at these successful applications of thermoelectric technology reveals the key reasons for this situation. An excellent and also very illustrative example is thermoelectric generators (i.e. TEGS) with a radioisotope for the heat source (i.e. Radioisotope Thermoelectric Generators, RTGs), which are widely used for electricity power supply in aerospace applications such as space craft, satellites and space probes for long-term-stable sources of electrical power (Champier 2017; Jaziri et al. 2020; Zoui et al. 2020). RTGs use the heat produced by the natural decay of radioactive materials and convert it into electricity. Hence, the characteristics of the heat source and the RTG are determined by the isotope material, which should preferably have low radiation emissions, an acceptable half-life within the mission's duration, a high melting point, a high power density and be safe in all conditions. The most suitable isotope fuel is the high-cost Pu-238, with a half-life of 89.6 years and heat energy of about 0.54 W/g (~ 5.2W/cm^3). The conversion efficiency of RTGs is up to about 8.5%, and they are capable of supplying electrical power up to about 300 W. Some of them have been operating for several decades, thus proving their reliability as a long-term source of electrical energy and thermoelectric devices in general. RTGs are the most important and specific commercial application of thermoelectric technology for energy production, despite their very high cost. The key basis for their successful application, however, lies in the fact that RTGs provide an unsurpassed solution for a long-term, non-stop, reliable supply of electrical energy without maintenance in a remote area (i.e. space, lighthouses, navigation beacons, weather stations, etc.), while price and efficiency are of secondary concern. The focus of the media and general public is of course on other aspects of space missions: technological aspects like power supply during the mission and the role of RTGs are mostly omitted. Another good example, also not in the public spotlight, are TEGs using the heat produced by the combustion of natural gas, butane or propane for power generation in industrial applications in remote areas, like deserts, well sites, offshore platforms and telecommunication sites in mountains (Champier 2017; Zoui et al. 2020). They can produce between 15 and 550 W, and can be combined for installations of up to 5,000 W. Their efficiency is low, around 2.3%; however, the high cost of energy is compensated by the fact that in these locations, power availability is critical and highly reliable power with low maintenance is required, regardless of the climate conditions.

Some thermoelectric products have been developed for general customers and are on the market. These products include camp stoves, designed for camping in general, but equipped with a thermoelectric generator using the heat of burning wood to produce electric power in the range from 2 W to about 10 W, depending on the design, for charging electrical devices via a USB port (Gao et al. 2016). On the Internet, the BioLite CampStove2+, weighing about 0.9 kg, with the boiling time for 1 liter of water being 4.5 minutes, enables charging with an electric power of 3 W and has 3,200 mAh of internal battery store. It sells for about $150. The possibility of charging the phone, navigation device and other electronic gadgets while cooking a meal in the wilderness is certainly very appealing, especially if we are intending to stay there for a longer time. And the price is basically affordable, hence, the commercial success of such a product depends primarily on its overall practical usability, as perceived by customers and confirmed in practice, which is the same as for most consumer goods. A camping stove with a charger for a mobile phone can be a practical solution in the wilderness, but not really significant in terms of energy sustainability. In contrast, it would certainly be a major step forward if we could harvest the waste heat from vehicles with a combustion engine, leading to significant savings in fuel. After all, two-thirds of fuel energy is lost as a heat, 40% of which is as a hot exhaust gas. If only about 6% of this exhaust heat could be converted into electrical energy, it would be possible to reduce fuel consumption by about 10%. Actually, tremendous progress has been made in this field, with various systems of thermoelectric generators (TEGs) that can be located either on the cooling system or on the exhaust system of cars and trucks (Jaziri et al. 2020; Zoui et al. 2020). TEG systems generating power from 10 W to about 610 W were reported, although their experimental conversion efficiency was below 2.6%. The technology of TEGs for vehicles with combustion engines is developed to the level of system prototype demonstration in the operating environment (i.e. TRL 7). There are still some technological difficulties and concerns related to the long-term durability, but the main obstacles for their broad application are basically economic reasons, expressed as a poor benefit-to-cost ratio. An additional element of uncertainty is the fact that electric cars are coming to the market at an accelerating pace, which could likely limit the use of TEGs to heavier vehicles with combustion engines, like trucks and ships.

In contrast to energy harvesting, thermoelectric cooling (TEC) applications are much wider spread. In comparison to conventional vapor-compression coolers with a high coefficient of performance (COP) from 2.6 to 3.0, TEC systems have low COPs of 0.38–0.45, but have numerous other advantages, which are essential to be able to ensure cooling in a variety of areas. Thus, TEC can provide solutions, and actually, in some cases, the only solution, to difficult thermal management problems where a low or moderate amount of heat is involved (Gurevich and Velazquez-Perez 2014; Zhao and Tan 2014; Atta 2018; Yu et al. 2019b; Baru and Bhatia 2020; Salah and Abuhelwa 2020). Some of the significant features of TEC technology are as follows: no moving parts due to the solid-state construction and hence no maintenance, high reliability with a very long life of about 23 years (compared to 10–12 years for conventional compressor coolers), small size and weight, scalability from less than 1 W of cooling up to several kilowatts, ability to cool below ambient temperature, accurate temperature control (better than ±0.1°C), silent operation, operation in any orientation and also in zero-gravity environments (i.e. space), convenient power supply (operated directly from a DC power source), spot cooling and environmentally friendly. In addition, it should be emphasized that the same TEC device can be used for both cooling and heating, and can also generate a small amount of DC electrical power when exposed to a temperature difference across it. Accordingly, TEC is used in many areas, for example: aerospace applications, avionics, black-box cooling, CCDs (Charge Coupler Devices), dehumidifiers, integrated-circuit cooling, laser-diode cooling, microprocessor cooling, electronic-package cooling, environmental analyzers, thermal cycling devices (DNA and blood analyzers) immersion coolers, IR detectors, night-vision equipment, and heated and cooled vehicle seats. A nice example is the TEC cooling of the Si(Li)-based detector in energy-dispersive X-ray spectrometers (EDS), replacing the conventional cooling by liquid nitrogen, which required the timely addition of nitrogen to the Dewar to ensure constant cooling, and preventing the destruction of the detector. This technical detail about the use of TEC is likely to be known to operators of modern scanning electron microscopes equipped with EDS detectors, and thus essentially limited to a specific group of scientific people. Unfortunately, the same applies for many other technical thermoelectric solutions. Another application is decorative wine-bottle coolers based on TEC, which could find a place in practically any home of those who like to enjoy a glass of good wine. They offer a silent operation, no shaking, accurate control of the temperature in just the right temperature range for cooling white and red wines, and low energy consumption. All advantages are difficult to overlook.

The above overview shows how extensive progress has been made with thermoelectric technologies in recent decades. They are well established in many areas, in particular where their unique advantages offer the best (and often the only) solution for particular needs, either in terms of energy supply or cooling, with the costs being of secondary concern. However, further broadening of their applications is hindered in many areas, primarily due to the low conversion efficiency and high costs. In general, two main roles of TEGs can be recognized. One is the thermoelectric recovery of waste heat from power plants (thermal, nuclear), industry, waste incinerators, automotive and other high-, medium- and low-power sources, which could contribute a significant amount of electrical energy. Here, the motivation is not so much the lack of energy, but more in the optimization of the energy efficiency. If there is no specific need for the produced electrical energy, the implementation of the thermoelectric harvesting predominantly depends on an economic justification. In the first place, the economics of such electricity harvesting depends on the costs of the thermoelectric technology, (i.e. the cost of the thermoelectric generators (TEGs) with the required infrastructure) and on the amount of produced electrical energy. The latter is directly dependent on the conversion efficiency of the TEGs. And the other, TEGs are a source of electrical energy for power supply. In this case, TEGs play a key role in enabling numerous low-power applications related to autonomous sensors, wireless sensor networks (WSNs), environment monitoring, medical applications, etc., thus also enabling IoT ecosystems.

Here, the focus is certainly on the conversion efficiency of TEGs and micro-TEGs to ensure a sufficient and reliable power supply from harvesting low- and very-low-power environmental heat sources into electricity.

The aim of this chapter is to provide a general overview of the current state of thermoelectric technologies, in particular with a focus on those aspects that have the main influence on their development, efficiency, cost and applications, i.e. thermoelectric materials, requirements for various applications of thermoelectric devices, principles of thermoelectric modules and their design and structure, and the fabrication technologies for TEGs. This overview will not be going into theoretical details with equations describing thermoelectric phenomena, the efficiency of materials and devices, the challenges in the development of thermoelectric materials with enhanced figure of merit (ZT) and the presentation of numerous thermoelectric devices for all ranges of envisaged applications, etc., which can be found elsewhere in numerous review articles on topics reported in the literature. In doing so, I hope to be able to expose some of the essential elements related to the problems and challenges of the complex development of thermoelectric devices for various applications, as well as prospects for the future.

7.2. Thermoelectric materials

Thermoelectric materials have the unique capacity of converting a flow of heat into electrical energy based on the Seebeck effect, and vice versa: by using an electrical current, the heat can be transferred from one place to another (Peltier effect). Exploitation of these phenomena requires the use of thermoelectric devices made from at least two materials with different Seebeck (S)/Peltier (Π) coefficients. The efficiency increases with the increasing difference in S (i.e. Π) between both involved materials. Hence, the use of n-type and p-type thermoelectric materials, preferably having a high absolute value of S (Π), one being negative and one positive, respectively, represents the best combination for a thermoelectric device. A basic thermoelectric pair is schematically presented in Figure 7.1. Thermoelectric modules typically contain between 10 and 100 of such couples of n-type and p-type legs, thermally connected in parallel and electrically in series. The thermoelectric efficiency of materials, and also thermoelectric devices, is generally expressed by the dimensionless figure of merit ZT, which is proportional to the electrical conductivity (σ), and the square of the Seebeck coefficient (S), and inversely proportional to the thermal conductivity (κ) (Snyder and Toberer 2008; Gould 2019; Zoui et al. 2020). In general, the physical properties of semiconductors provide the best way to meet the requirements for such contradictory parameters for good n- and p-type thermoelectric materials. The relationship between the value of ZT and the conversion efficiency is shown in Figure 7.2. As with the other "heat engines", also for thermoelectric devices, the maximum theoretical efficiency is limited by the Carnot efficiency. Traditional power generators can reach 40% of Carnot efficiency, and for such conversion efficiency, the ZT value of the thermoelectric materials should be at least 3. Accordingly, a lot of effort has gone into the development of new thermoelectric materials with a high ZT and to enhance the ZT of already-known thermoelectric materials in recent decades, after the advantages of thermoelectric technologies were widely recognized.

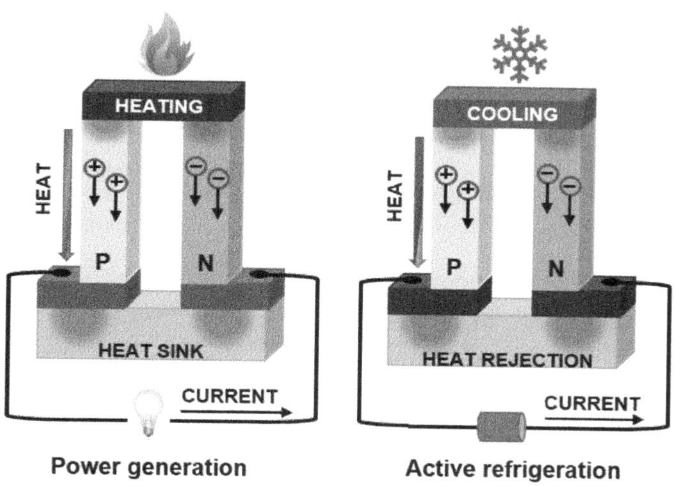

Figure 7.1. *Schematic of heat conversion into electricity (i.e. power generation mode – Seebeck effect) and thermoelectric cooling (i.e. active refrigeration mode – Peltier effect)*

Thermoelectric materials are often classified into two categories, i.e. conventional and novel materials. The conventional thermoelectric materials, i.e. Bi_2Te_3/Sb_2Te_3, PbTe, TAGS and Si–Ge are mostly based on heavy elements like Te, Bi, Pb, Sb and Ag, which are rare, expensive, heavy and toxic. These materials have been predominantly used in hitherto successful applications of thermoelectric generators (i.e. RTGs in space programs) and for cooling, mostly limited to niche areas of technology. The current thermoelectric technologies are still largely based on these conventional materials with ZT values around 1. Some materials from this group are still irreplaceable, especially those with good thermoelectric characteristics for use at temperatures below 150°C. In particular, Bi_2Te_3 has superior thermoelectric performance in this temperature range with a ZT value of 1 at room temperature and has no real alternative, despite long-term efforts to replace it. Hence, Bi_2Te_3 is nowadays mostly used in the development of body-heat harvesting applications and others related to environmental heat sources with low power and small temperature differences (Li et al. 2019; Nozariasbmarz et al. 2020a, 2020b, 2019a). The other materials from this group, like TAGS and PbTe-based materials, are suitable for applications in an intermediate temperature range of 150°C to 500°C. The Si,Ge materials are considered suitable for temperatures above 500°C. In contrast to the other classified materials, Si,Ge is made from light, abundant, cheap and environmentally benign elements, which are very useful characteristics for many novel thermoelectric materials.

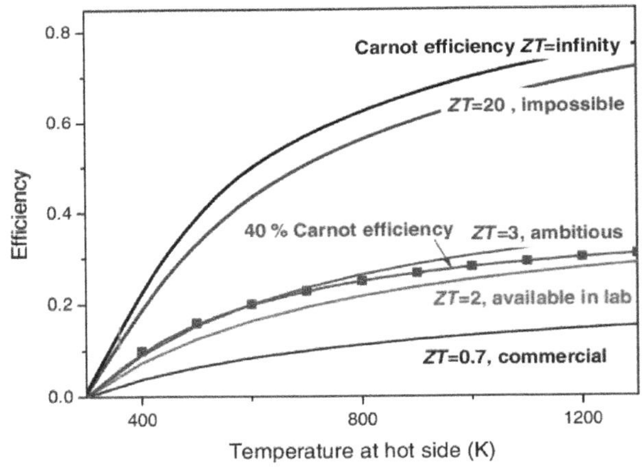

Figure 7.2. *Conversion efficiency of TE materials depending on the temperature difference and ZT, assuming that the cold side temperature is 300 K (Han et al. 2014)*

The family of novel thermoelectric materials is very diverse (Aswal et al. 2016; Masood et al. 2018; Cai et al. 2019; Freer and Powell 2020), comprising materials such as skutterudites (Rogl et al. 2014; Rogl and Rogl 2017; Liu et al. 2020), clathrates (Christensen et al. 2010; Takabatake et al.

2014; Sui and Kauzlarich 2016), half-Heusler alloys (Chen and Ren 2013; Poon 2018; Ciesielski et al. 2020), chalcogenides, i.e. sulfides and selenides (Chetty et al. 2015; Li et al. 2018; Wei et al. 2018; Shi et al. 2019; Powell 2019; Qin et al. 2020; Yu et al. 2020; Guélou et al. 2021), silicides (Liu et al. 2017; Nozariasbmarz et al. 2017; Burkov 2018; Kim et al. 2019, 2020) and oxides (Feng et al. 2017; Ji 2018; Bernik 2021). Based on their ZT values and chemical stability, most of them are suitable for applications in the medium-temperature range from 250 to 650°C, while some half-Heusler alloys, silicides and in particular oxides are also suitable for temperatures above 650°C. Their development in recent decades was strongly affected by the concept of phonon-glass electron-crystal (Rowe 1995) and nanostructure engineering (Chen et al. 2012; He et al. 2013; Mohanraman et al. 2015), leading to an increase in ZT to 1 and above. Nowadays, quite a few novel thermoelectric materials have ZT values between 1 and 2 and some even above 2 (Rull-Bravo et al. 2015; Aswal et al. 2016; Cai et al. 2019; Yu et al. 2019a; Shi et al. 2020). It is certainly good that in all these groups (i.e. systems) of novel thermoelectric materials, the p- and n-type materials are available; unfortunately, their thermoelectric characteristics within the same group mostly do not match, as one type of material (i.e. usually the p-type) has a distinctly higher ZT value than the other. Research efforts are therefore often used to enhance the thermoelectric characteristics of this inferior type of material from the group, to have p- and n-type materials with the highest possible and similar ZT values in the same system. In the case of skutterudites, however, n- and p-type materials with comparably high ZT values in the range from 1 to 2 are already available (Rull-Bravo et al. 2015; Rogl and Rogl 2017). Hence, skutterudites are considered as very promising candidates for medium-temperature thermoelectric power generation. The ZT values for some state-of-the-art thermoelectric materials are presented in Figure 7.3. The commercially available materials have mostly ZT values of about 0.7; it is considered that such an average value of ZT for a material under an applied temperature gradient would be sufficient for energy harvesting. However, looking generally and not just at current ZT values, each group of novel thermoelectric materials has some interesting advantages for applications in comparison to the others, which encourages and justifies their further development. This could be environmentally friendly composition (sulfides, silicide, oxides), made from abundant and low-cost elements (sulfides, silicides, oxides), lightweight (silicides), stability in air at moderate (skutterudites, chalcogenides) and high (half-Heusler alloys, silicides, oxides) temperatures, etc.

When referring to novel thermoelectric materials and their progress, we should certainly not forget two groups of materials which will likely have an immense impact on thermoelectric technologies, particularly at room temperature. One is silicon, which has long been considered unsuitable as a thermoelectric material due to its high thermal conductivity. This changed when a discovery was made: by reducing the dimensions to submicron sizes, the thermal conductivity of silicon is drastically reduced (Boukai et al. 2008; Hochbaum et al. 2008). The development of thermoelectric silicon that followed was a strong motivation, with the possibility to facilitate the integration of thermoelectric technologies with Si-based semiconducting electronics, which is the core of our electronics and information technologies. This resulted in doped silicon materials with ZT values of up to 0.6 at room temperature (Tang et al. 2010; He et al. 2019; Tanusilp and Kurosaki 2019; Pham et al. 2020).

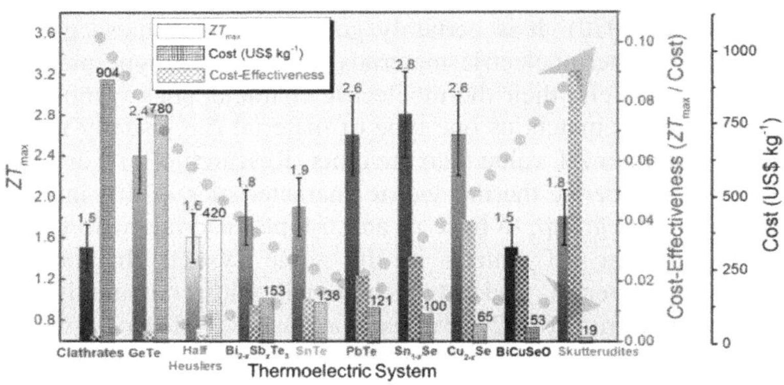

Figure 7.3. *ZT values of state-of-the-art thermoelectric materials in the last 10 years and their costs (Shi et al. 2020)*

The other group are organic thermoelectric materials (Zhang et al. 2014; Russ et al. 2016; Chakraborty et al. 2018; Wang and Yu 2019; Zhang and Park 2019; Zhang et al. 2019; Lindorf et al. 2020) represented by the p- and n-type conjugated polymers (Peng et al. 2017; Yao et al. 2018, 2019), carbon-nanotube-based materials (CNT) (Blackburn et al. 2018; Wang et al. 2018; Hung et al. 2019) and graphene-based materials (Amollo et al. 2017; Zong et al. 2020). Their development resulted in peak ZT values up to 0.5. They are mechanically flexible, lightweight, cheap, environmentally friendly, easily fabricated in versatile formats and thus very suitable for thermoelectric energy-harvesting applications in wearable electronics, body-heat energy harvesting and in general for applications at environmental temperatures.

The focus of the research in recent decades and to a large extent even now is on improving the value of ZT; after all, conversion efficiency and with it the output power of thermoelectric devices are positively correlated with the ZT of materials. However, ZT is a highly temperature-dependent parameter and the conversion efficiency is also higher with higher operating temperature differences, rising almost linearly with ΔT, as is clearly evident in Figure 7.2. Unfortunately, for most thermoelectric materials, the maximum ZT values relate to a relatively narrow temperature range of about 200–300 K. Hence, for better efficiency in exploiting high-temperature heat sources, it is desirable to optimize the ZT over a large temperature gradient, which can be done using the concept of segmented thermoelectric modules (Bai et al. 2021). It means that selected thermoelectric materials with different optimum operating temperatures along the direction of the temperature gradient are combined in an electrical series to form each of the p- and n-type thermoelectric legs. It soon showed, however, that a simple choice according to suitable ZT values is not enough and that materials combined into a segmented device must be thermoelectrically compatible to achieve an improved efficiency; if they are not, the efficiency will markedly decline in comparison to that expected. Based on an understanding of the thermodynamics behind segmented thermoelectric devices, the thermoelectric compatibility factor (s) was introduced (Snyder and Ursell 2003), defined as $s = (\sqrt{(1+ZT)} - 1)/ST$. In accordance with their report, the maximum efficiency that the thermoelectric material can provide is only achieved when the relative current density (u, i.e. the ratio of the electrical current density to the heat flux) is equal to s. This means that for an effective thermoelectric device, the u value has to be roughly constant through a segmented element, changing by less than 20%. A difference in the compatibility factor by a factor of about 2 or more means that a given u cannot be suitable for materials connected in series and the segmentation will not be efficient. Hence, materials with a high ZT are needed, which also have similar s. Given this, continuous efforts to develop a variety of p- and n-type materials with enhanced thermoelectric properties in order to have a good choice of candidates for constructing the segmented modules according to optimal ZT and compatibility are even more understandable.

ZT and s are certainly very important in segmented devices, because the thermoelectric properties can change dramatically from one segment to another. However, there are also many other chemical and physical properties affecting the selection of materials, which are not important only for segmented, but actually in general for any thermoelectric module and were somehow in the shadow of headline efforts to improve the ZT of materials and their conversion efficiency. Due to the service conditions of TEGs, these materials are unavoidably subjected to different temperatures and a wide range of temperature gradients created within the materials themselves. Hence, the temperature dependence of any property of the thermoelectric material is a subject of concern. These other property factors are the thermal and chemical stability, mechanical and thermomechanical properties, aging and fatigue; they affect the long-term reliability of materials and modules. Their influence is becoming increasingly critical with an increasing operating temperature and temperature gradient across the thermoelectric module, and also due to the frequency of the temperature cycles of heating and cooling to which the module is exposed. The Young's modulus (i.e. the elastic response to uniaxial loading, either tension or compression), strength (i.e. the stress that a material can withstand under any load without a permanent change), hardness (i.e. the material's resistance to penetration, also indicative of a resistance to wear) and the fracture toughness (i.e. the ability to withstand loading by plastic deformation before catastrophic failure; a brittle material is without an ability to plastically deform) are the most crucial mechanical parameters for designing devices and have to be correct across the entire operating temperature range. After all, these mechanical properties of thermoelectric materials are important for their ability to be machined into the shapes needed for the manufacturing of modules. They are also important for the physical stability, robustness and structural integrity of the modules, including normal handling, and also under operating conditions for any thermal gradient and possible mechanical loads (i.e. compression, vibrations, etc.). In operating conditions, the thermal shock resistance (i.e. the material's ability to withstand the thermal stress introduced by a sudden increase in temperature) and the fatigue resistance (i.e. the capability of the material to withstand multiple cycles of heating and cooling) also must be taken into consideration. The thermal stress ($\sigma_{thermal}$) developed in the material is proportional to the coefficient of thermal expansion (α), the Young's modulus (E) and the temperature difference (ΔT) across the material. As long as the thermal stress is lower than the fracture stress (fracture toughness), the material will withstand the shock. It is also relevant for the material to have good (i.e. higher) thermal fatigue resistance, which means that it is capable of withstanding a larger number of thermal cycles. In recent reports, the problem of the relevant mechanical properties for practical applications and the mechanical characteristics of some thermoelectric materials (i.e. (Bi/Pb)-tellurides, skutterudites, half-Heusler alloys, (Si,Ge)-based alloys) has been well presented (He et al. 2015; Wereszczak and Case 2015; Guttmann and Gelbstein 2018). According to these reports, most of the thermoelectric materials exhibit an elastic modulus similar to common engineering materials in the range from several tens of GPa up to about 200 GPa, which is encouraging for the design of thermoelectric devices. Unfortunately, thermoelectric materials are mostly brittle and show no yield strength. Therefore, their fracture tends to be sudden and catastrophic. The hardness of thermoelectric materials can vary a great deal: from values much below and around 1 GPa to about 14 GPa (Kenfaui et al. 2010; He et al. 2015; Guttmann and Gelbstein 2018). For the fracture toughness (in MPa $\times$ $m^{1/2}$) of the thermoelectric materials, values in the range of about 0.3 to about 2.8 are reported (Guttmann and Gelbstein 2018, Li et al. 2021). These mechanical properties are valid for room temperature; however, it would certainly be beneficial to also know their temperature dependence in the temperature range of the intended use. Thermoelectric materials are brittle in nature and prone to damage by thermal shock. While the thermal shock resistance can be evaluated via other mechanical properties (Wang et al. 2016; Guttmann and Gelbstein 2018; Wang and Li 2018), there is certainly a lack of studies about their thermal fatigue resistance. The mechanical properties of materials are determined by their chemical composition and crystal structure, with the nature of the bonds between the atoms and the microstructure (i.e. the phase composition and the distribution of phases, the morphology of the grains, the grain size and size distribution, texturing, porosity, etc.). Hence, they can be greatly affected by the processing of materials due to the rearrangement of the internal structure (i.e. microstructure), which gives an opportunity for their enhancement. In general, the use of p- and n-type materials from the same family, which are chemically and structurally as close as possible, can also mean similar mechanical and chemical properties, and therefore not only better mechanical but also chemical compatibility in the module. In this regard, another factor that has to be considered, in particular for an increasing operating temperature to which the materials are exposed, is the thermal expansion. Thermal expansion means the tendency of a material to change its shape, area, volume and density as a response to a change in temperature, which is described by the thermal expansion coefficient, linear, areal or volumetric, and usually does not include phase transitions. The choice of coefficient depends on the particular application and which dimensions are considered as important. The thermal expansion is closely related to the nature of the bonding in materials and decreases with an increase of the bond energy. Hence materials with a higher melting temperature are more likely to have a lower thermal expansion coefficient. If the body is free to expand, the expansion or strain resulting from an increase in temperature can be simply calculated using the applicable coefficient of thermal expansion. In a module, however, the materials are not free standing and their expansion is restrained. In this case, the stress resulting in a material can be calculated by considering the strain during free expansion and the stress required to reduce that strain to zero, using the stress/strain relationship characterized by the Young's (elastic) modulus. Thermal expansion is certainly a factor threatening the structural integrity of modules and can cause their degradation, especially if thermal cycling is applied, which makes the thermal expansion coefficient a key design parameter for thermoelectric materials (Ravi et al. 2009; Ni et al. 2013; Shtern et al. 2019). Accordingly, it is good if a selection of materials, compatible in all the required respects, is available and with coefficients of thermal expansion that do not vary significantly over the range of temperatures where they are designed to function. A nice example is the large family of p- and n-type skutterudites with a linear thermal expansion coefficient α somewhere in the range from about $5\times10^{-6}K^{-1}$ to about $14\times10^{-6}K^{-1}$ (Schierning et al. 2015). In general, n-type Sb-based skutterudites have a lower α than a p-type isotype; however, for some Fe/Ni-substituted multi-filled skutterudites, a p- and n-type material with practically the same α and ZT of about 1 can be produced, also having identical hardness and elastic modulus.

For applications it is also significant that thermoelectric materials are available in sufficient quantities of constant and reproducible quality, and for a low price. However, in the development of thermoelectric materials, very small amounts are typically prepared, sufficient only for their characterization. Hence, until quite recently, many materials with good thermoelectric characteristics were only available in small amounts, often below 1 gram. In general, the processing of bulk thermoelectric materials mostly consists of two stages, i.e. the synthesis of the material in the form of powder or bulk, and the final consolidation of the material by sintering. Processing is well known to affect the thermoelectric characteristics of materials significantly and is therefore a powerful tool that has been widely studied and is already well exploited to enhance them (Recatala-Gomez et al. 2020; Ma et al. 2021). But processing, and in particular synthetic methods, also have an important influence on the possibility of scaling up the production of the material from the laboratory to the semi-industrial and industrial quantities needed for applications. Besides the conventional methods of solid-state synthesis and melting, in recent years, novel thermoelectric materials (i.e. chalcogenides, clathrates, skutterudites, half-Heusler alloys, silicides and oxides) were developed using different synthesis methods, such as wet chemical methods (i.e. solvothermal and hydrothermal methods, co-precipitation, etc.) and mechanochemical alloying (i.e. high-energy milling); they enable the reproducible preparation of materials and are relatively easy to scale up. Many of these materials, with their excellent thermoelectric characteristics, are now available in much larger amounts because they can be reproducibly synthesized in batches of several 100 grams. Also worth noting is that these synthesis methods, which are usually combined with hot-pressing and spark plasma sintering (SPS), inherently support nanostructuring and nanostructural engineering for enhanced thermoelectric characteristics (Bux et al. 2011; Baláž et al. 2017; Li et al. 2017; Wu et al. 2017; Masood et al. 2018; Baláž et al. 2019, 2020; Lei et al. 2019; Young and Reddy 2019; Hamawandi et al. 2020; Kruszewski et al. 2021; Novitskii et al. 2021). As these novel thermoelectric materials are mostly made from abundant and hence cheap chemical elements, it is certainly desirable for them to be processed by cost-effective methods. However, for a more comprehensive review of the economics of certain thermoelectric materials, in addition to the price of materials, their thermoelectric efficiency (ZT) should also be taken into consideration (see Figure 7.3); a cheaper material with a higher thermoelectric conversion efficiency means a lower price per watt of produced energy (LeBlanc 2014; LeBlanc et al. 2014; Shi et al. 2020; Jaldurgam et al. 2021). The two-stage processing of bulk thermoelectric materials for standard Π-type thermoelectric modules is nevertheless time- and energy-consuming, which certainly add to the production costs. Hence, rapid fabrication methods were also developed in recent years, such as the laser-melting, additive-manufacturing technique (El-Desouky et al. 2016).

7.3. Thermoelectric devices – structure, materials, fabrication technology

In general, the term thermoelectric devices includes devices for thermoelectric energy harvesting and thermoelectric cooling. Here, primarily, thermoelectric energy-harvesting devices will be discussed, although most of the content could also apply to thermoelectric cooling devices. Structurally, the main part of thermoelectric modules is the thermopile, which consists of thermocouples (Figures 7.1 and 7.4) connected electrically in series and thermally in parallel. When the thermoelectric module is connected to a heat source on one side and a heat sink on the other, we are talking about a thermoelectric generator (TEG). For the optimum conversion efficiency of the TEG, the good transfer of heat from the heat source to the module is essential, as well as the highest possible and constantly maintained temperature gradient across the module by the good transfer of heat from the module to the heat sink. The standard Π-type sandwich structure of the module is shown in Figure 7.4c. This is the traditional structure of thermoelectric modules and is the most commonly used. It is typical of thermoelectric devices used for the generation of electric power in extreme environments, like RTGs for space explorations, TEGs for industrial applications in remote areas and TEGs for the recovery of waste heat for the production of electricity (Champier 2017). Actually, it is widely adopted in commercial thermoelectric products; the thermoelectric modules, which can be found on the Internet and are available on the market, have such a structure. The standard Π-type structure is characterized by vertical heat and current flows (also called a vertical structure), while in comparison to other structures, its main advantages are a simple design and the fact that it enables a large power output and a high conversion efficiency. In power-generation applications, these are certainly advantages that outweigh the main drawbacks, which are the difficult and time-consuming fabrication and the high-technology requirements. The vertical structure is also widely used in TEGs for micro-energy harvesting (μTEGs). However, in micro-energy harvesting, certain advantages of other structures, i.e. lateral and hybrid structures, are also required (Yan et al. 2018). All the different types of structure, i.e. vertical, lateral and hybrid, are schematically presented in Figure 7.4. In the lateral structure, the heat and current both flow in the lateral direction. The main advantage of the lateral structure is that it is easily fabricated in μ-TEG for its compatibility with integrated-circuit (IC) planar technology. The main disadvantage is its comparatively poor performance due to a high parasitic heat flux through the membrane and the difficulty in packaging this chip-like device to apply an in-plane temperature difference. Hence, it is rarely used in power generation. The lateral structure is normally used in different sensor applications. The hybrid structure combines the performance of the vertical and lateral types of structures and thus integrates their advantages; the heat from environment is provided vertically into one side of the chip, guided through the thermopile in a planar direction and ejected vertically on the other side. The packaging of thermoelectric modules with a hybrid structure is convenient due to the vertical direction of the heat flux on both sides, as in the case of the vertical structure. Furthermore, the in-plane thermopiles can be fabricated using

IC-compatible technology, as in the case of the lateral structure. The main challenge in the design of the hybrid structure is the optimization of the heat-flux path, which is essential for improving the output power of such a TE module. The possible effective solutions include introducing a cavity beneath or above the thermopile, a metal layer on both sides of the module to improve its thermal coupling with the surrounding environment, and the optimization of the shape and the distribution of thermopile. The hybrid structure is certainly desirable when its performance complies with particular applications (Yan et al. 2018).

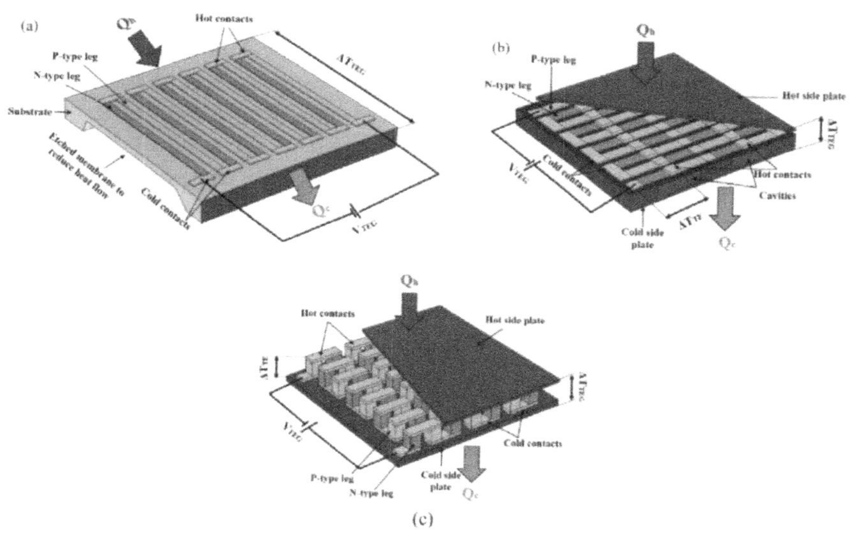

Figure 7.4. *Schematic of structures used in the fabrication of micro thermoelectric devices (μTE-modules) and the directions of the heat and current fluxes; a) lateral, b) hybrid and c) vertical (i.e. standard Π-type) (Jeziri et al. 2020)*

However, within a given structure, the efficiency and hence the power and voltage outputs can be further optimized using the geometrical parameters, which make it possible to adjust the differences in the electrical and thermal conductivities between the p and n legs of the thermocouples and thus affect the current and heat transport through the device. The main geometrical parameters are the leg length, the aspect ratio (i.e. ratio of the leg length to the leg cross-sectional area) and the fill factor (i.e. the proportion of a thermoelectric device's area, which is occupied by the thermoelectric material). Accordingly, for different electrical and thermal conductivities of the p- and n-type legs of the thermocouple, the current and heat flow through the legs can be equilibrated for maximum efficiency via the geometry of the legs, usually by adjusting the cross-section of each leg, because the lengths of the thermocouple legs are usually equal to avoid the complication of the design and fabrication process, i.e. a larger cross-sectional area of the leg from the material with the higher resistivity. In the case of the lateral and hybrid structures of the thermoelectric module, the thickness of the thermoelectric material is typically limited by the fabrication technology, and so the cross-section is commonly adjusted by the width of the leg. In general, when a high-power source of heat is available, then the design with a small aspect ratio of short legs can be used to maximize the power generation. However, this is at the expense of the efficiency. In contrast, a high-aspect-ratio design can be used in the case of limited heat energy to maximize the efficiency at the expense of the power generation (Kim et al. 2016; Yan et al. 2018; Thimont and LeBlanc 2019; Freer and Powell 2020). The fill factor is reflected in the thermal resistance parameters of the module; it affects the amount of heat transfer from the hot to the cold side and hence the temperature gradient in the module. With a smaller fill factor, the optimum thickness for the maximum power output is lower (Bahk et al. 2015; Zhu et al. 2020). The geometrical parameters also affect the amount of thermoelectric materials needed in thermoelectric module and hence also the costs. In the case of more expensive materials, reducing the volume of the legs while keeping the same aspect ratio can reduce costs. However, it must be considered that this affects the thermal and electrical resistances and hence the output power and efficiency of the module. The geometry parameters of thermoelectric module greatly affect its performance (i.e. the efficiency and power output). Hence, the geometrical design and its optimization are essential for the successful development and manufacturing of thermoelectric devices, while modeling and simulations can also help (Lee 2013; Meng et al. 2016; Ma et al. 2017; Ferreira-Teixeira and Pereira 2018; Dongxu et al. 2019; Karana and Sahoo 2019; Wolf et al. 2020; Tanwar et al. 2021).

The overall complexity of thermoelectric devices and the engineering challenges they present are only fully revealed once it comes to their assembly and manufacturing, which actually determine their practical energy-conversion efficiency and service life. Aside from the thermoelectric materials with a direct ability for energy conversion expressed by ZT, which have so far largely been in the focus of research and prevail in literature reports, thermoelectric devices comprise several other materials, each with a specific function in their structure. So, the integration of thermoelectric conversion materials into a thermoelectric device also requires the following key materials: i) thermoelectric interface materials, ii) package materials and iii) housing materials. In addition, solder or brazing materials and electrode stripes are necessary for a thermoelectric module (Liu and Bai 2019). The key materials for assembling thermoelectric devices are schematically shown in Figure 7.5. Their specific characteristics required in regard to their role in the structure itself, very much affect the possible choices and bring engineering and cost challenges with respect to their availability, compatibility and price. The required characteristics of thermoelectric conversion materials were presented in the previous section. When it comes to their connection with other materials in a thermoelectric device, the following are especially relevant: high electrical and low thermal conductivities, a proper mechanical strength and the thermal expansion coefficient. Interface materials connect the p- and n-legs of thermocouples with an electrode strip. Hence, they must have high electrical and high thermal conductivities to support the low-energy-loss transport of heat and electricity through the device, and must also bond easily and well to the thermoelectric conversion material. Good packaging materials have low electrical and thermal conductivities and high thermal stabilities. Packaging materials fill the empty space between the columns of thermocouples within the structure of the thermoelectric module, which is especially important in the case of a very low fill factor, when it also has to provide mechanical support. Packaging materials encapsulate the thermoelectric conversion materials and thus protect them from environmental (i.e. atmospheric) influences. Housing materials need to have a low electrical conductivity to electrically insulate the module from its surroundings and a high thermal conductivity for the good transfer of heat from the heat source to the module, and on the cold side, good transfer of heat from the module to the heat sink. The housing material has to be thermally stable and mechanically strong to provide mechanical integrity to the module and for the protection of the thermoelectric conversion material. The choice of housing materials is rather limited, in power generation mostly to ceramic substrates of Al_2O_3, AlN, Si_3N_4 and others, which are expensive.

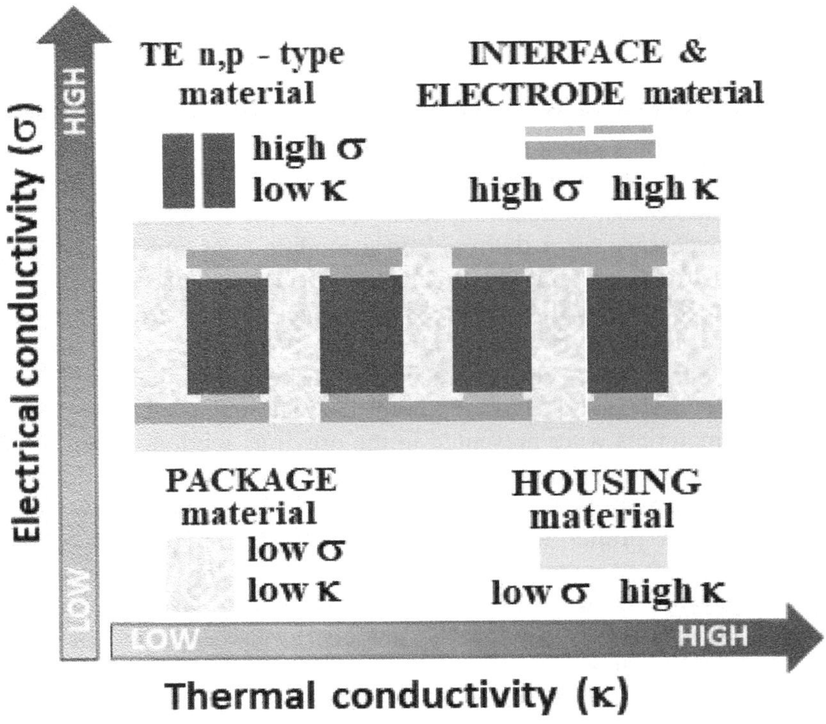

Figure 7.5. *Schematical presentation of key materials for thermoelectric power generation device: thermoelectric n- and p-type materials, interface and electrode materials, package material and housing material*

A variety of materials, interconnected in thermoelectric modules and hence with very different interfaces, means real, open engineering challenges in regard to bonding strength, contact resistivities, as well as thermal and chemical stability. In this regard, in particular the connection of the interface material with the thermoelectric conversion material is important as it directly affects the electrical and thermal resistivities of the module, hence the transport of heat and electricity, and consequently the conversion efficiency. The challenges of the key interfaces in thermoelectric modules, and the connections between the interface requirements and the design principles are schematically shown in Figure 7.6. The problem of the interfaces in thermoelectric power-generation modules has been thoroughly discussed in recent articles (Liu et al. 2015; Liu and Bai 2019).

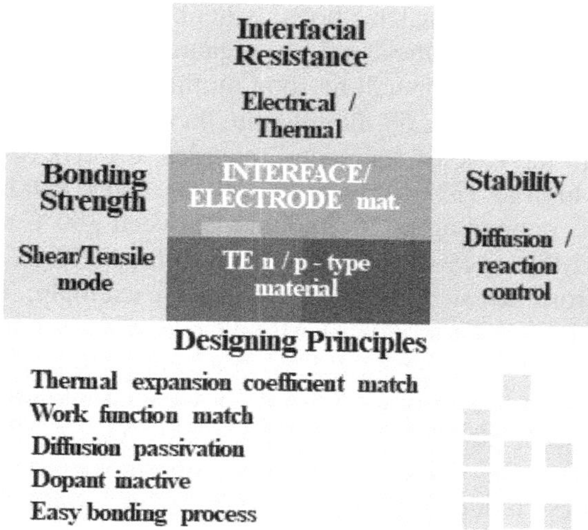

Figure 7.6. *Schematic presentation of the challenges related to key interfaces in thermoelectric modules, and the connections between the interface requirements and the design principles*

The development of thermoelectric modules has been stimulated in recent years by the requirements for sustainable technological development, the needs for an alternative supply of power in future advanced technologies and also significant progress in the development of new thermoelectric materials based on elements that are abundantly available in nature, cheap and environmentally friendly, with enhanced thermoelectric properties. Accordingly, the development of thermoelectric devices also greatly expanded the recovery of waste heat in industries and transport into electricity, and to the scavenging of low-power environmental heat sources and human body heat into electricity for powering sensors, wireless-sensor technologies and IoT, microelectronics, wearable electronics, medical devices, etc. A roadmap for the exploitation of thermoelectric technology has recently been proposed (Freer and Powell 2020).

The applications of thermoelectric devices in power generation are still mostly limited to the generation of electricity in extreme environments with RTGs for space exploration and TEGs for industrial applications in remote areas (Champier 2017; Zoui et al. 2020). However, in recent years, a number of thermoelectric modules for low- (<250°C), mid- (250–650°C) and high-temperature (>650°C) ranges, based on novel thermoelectric materials (i.e. skutterudites, half-Heusler alloys, silicon-based alloys, Mg-silicide, tetrahedrites, calcium/manganese oxides, organics) have been developed and some are already available or will be soon on the market (Aswal et al. 2016; Champier 2017). Depending on the module, their efficiencies are between 1 and 8.5%, and in the case of segmental modules, the reported experimental efficiency is as high as 15% (Aswal et al. 2016). Their maximum operating temperatures are, depending on the module, between 300°C and 800°C, and at temperature differences in the range from 300 to 750 K, their produced power ranges from about 3 W to 25 W. As an example, TECTEG MFR calcium/magnesium oxide modules CMO-32-62S with an upper temperature limit at 850°C can produce up to 12.3W in a temperature gradient of 750K. At a size of 64.5mm × 64.5mm × 8.6mm, they contain 32 couples, and their price in 2015 was $360. The same producer also put on the market, for a price of $560, the first segmented thermoelectric modules of calcium/manganese oxide bonded with Bi_2Te_3 on the cold side. Their size is the same as the CMO-32-62S and can produce 11 W at an applied gradient of 435°C (assuming a 45°C cold-side temperature). Also, the skutterudite-based TEG25 made by the Shanghai Institute of Ceramics can produce up to 25W at a temperature gradient of 510 K (cold-side temperature 65°C). The dimensions of the module are 50mm × 50mm × 10 mm, and it contains 32 couples of 5 mm × 5 mm × 8 mm (Champier 2017).

Micro-energy harvesting with thermoelectric modules for the electric power supply of low-power devices such as sensors, micro-electronics, biomedical devices, etc., would enable their autonomy and facilitate the spread of applications for many advanced technologies with a strong impact on the quality of life, including the concept of the IoT (Akinaga 2020). Exploiting environmental heat sources certainly represents a great challenge as the energy density of these sources is very low, somewhere between 1 and 1,000 $\mu W/cm^2$ (or $\mu W/cm^3$) and the temperature gradients on the module are mostly much below 50K and are often only a few degrees (Zabek and Morini 2019). In the case of the human body, the amount of heat varies from about 27°C to 35°C, depending on body's activity, the location on the body and the environmental conditions, while the average temperature difference between the body and the ambient air is around 13°C (Kumar et al. 2019; Jaziri et al. 2020). It might seem trivial, but maintaining a temperature gradient in such operating conditions, especially at temperatures around room temperature, is very demanding and among the key issues for the efficiency of μTEGs. In recent years, numerous thermoelectric devices for micro-energy harvesting have been developed at the prototype stage and presented in articles with their parameters relating to structure (vertical, lateral or hybrid), material, performance, etc. (Yan et al. 2018; Jaziri et al. 2020). In the thermoelectric harvesting of body heat for the electricity supply to wearable electronics and for biomedical applications, flexible thermoelectric modules are often desired and have been developed (Rojas et al. 2017; Kumar et al. 2019; Jaziri et al. 2020; Nozariasbmarz et al. 2020b; Fernández-Yáñez et al. 2021). In comparison to traditional thermoelectric modules, these micro-devices have the advantages of small size, light weight, low cost and high voltage factor, which are characteristics desired in the fields of sensor technologies, consumer electronics, biomedical engineering and the IoT. However, they are still predominantly based on Bi_2Te_3, although other novel materials are increasingly being used. It is also important for their development and for their compatibility with the planar technology of integrated circuits and electronics that they can be fabricated using technologies that are already well established in the industry of advanced electronics and electronic devices, including thin-film deposition methods (i.e. PVD, CVD, PLD, magnetron sputtering, flash evaporation, electro-chemical deposition), thick-film printing methods (i.e. screen printing, 3D printing) and in the technologies of MEMS, CMOS and CMOS-MEMS (Orrill and LeBlanc 2017; Yan et al. 2018; Jaziri et al. 2020).

It is difficult to compare the performance of the reported μTEGs due to the lack of standard measurements; however, their output power is for now typically somewhere in the range from μW to mW. For instance, the power requirements of sensors for measurement are in the range from about 3 μW (i.e. humidity sensor) to about 2 mW, and in standby mode from about 0.25 μW to about 17 μW. In the case of wireless sensors and wireless-sensor networks (WSNs), the requirements of power consumption for transmit/receive are from about 40 to 400 mW (Tang et al. 2018). In medical applications, the power requirements for common implantable devices (i.e. cardia pacemaker, drug pump, glucose monitor, etc.) are in the range from about 20 μW to about 10 mW (Kumar et al. 2019). Accordingly, μTEGs are still somewhat on the short in terms of generated output power for most of the intended applications. The low output power, which cannot meet the requirements for most WSNs, is a major drawback for the widespread use of μTEGs in the IoT. The main reasons for the low output power are the low ZT of the thermoelectric material, too high contact resistances within the thermoelectric modules (in particular between the thermoelectric semiconductor and the metal of the interface and the electrode) and the low thermal resistance of the module from the hot side to the cold side. These are also the main challenges that will have to be overcome to improve the performance of thermoelectric micro-harvesting devices to meet the requirements of applications.

7.4. Summary

Thermoelectric materials and devices have undergone intense development and tremendous progress over previous decades, as the successful implementation of thermoelectric technologies would also enable the successful application of many advanced technologies on a large scale, including efficient cooling for special requirements, harvesting waste heat into electricity and the energy autonomy of various systems for the IoT, to list just some of the options. The future is in electricity, but about 60% of all primary energy in the world is being wasted as heat; converting only a fraction of this heat into electricity would be a huge benefit. Thermoelectric technology is the only one that can convert heat into electricity directly, regardless of the power of source, without moving parts, without the need for maintenance and without environmentally hazardous side products. The reliability of thermoelectric technology is confirmed by the decades of stable operation of radioisotope thermoelectric generators (RTGs) as power sources on space missions and in the industrial applications of thermoelectric generators (TEGs) in remote areas.

Today, we have many novel n- and p-type thermoelectric materials based on abundantly available, cheap and environmentally friendly elements, which have excellent thermoelectric properties at higher temperatures with ZT values of around 1 and up to 2, and some even higher. It is important that some of them, such as skutterudites and chalcogenides (sulfides, selenides, tetrahedrites), are already available in larger quantities as they can be produced by processes which allow for reproducible qualities and an easy scaling up of the production process. Based on these novel materials, various thermoelectric devices have been developed in recent years at a technology readiness level (TRL) 3 to 5 (i.e. the laboratory prototype stage) such as thermoelectric modules for the recovery of the waste heat into electricity and also rigid and flexible modules with different structures (vertical or standard p-type, lateral and hybrid) for the micro-energy harvesting of electricity based on the small environmental temperature differences as well as human body heat. It is certainly an advantage that these thermoelectric modules are developed and fabricated using advanced technologies which are already well established in the industry of electronics and electronics devices, including thin- and thick-film deposition methods, and the technologies of MEMS, CMOS and CMOS-MEMS.

Nevertheless, despite significant progress in thermoelectric materials and devices, the main limitations for the wider use and commercialization of thermoelectric technologies are the still too low conversion efficiency and high manufacturing costs. Hence, the further development of thermoelectric materials is essential to ensure the supply in sufficient amounts of high-quality thermoelectric materials with high ZT values across a broad temperature range and for an appropriate (i.e. low) price. Certainly, thermoelectric materials with high conversion efficiency at temperatures below 300°C would be needed, where the most commercial applications are envisaged. Also, the efficiency of thermoelectric modules will have to be improved significantly and their manufacturing costs must be reduced. This requires the use of low-cost thermoelectric materials with high ZT values, which are also thermoelectrically compatible, the optimization of all materials assembled in the module with regard to their thermo-electrical, thermo-chemical and thermo-mechanical compatibility, and also the further enhancement of their fabrication technology, all with the scope to minimize the resistivities of the interfaces in the module that are the main reason for the low efficiency, and to improve the long-term stability of the modules. In doing so, the use of modeling can be a powerful tool for optimizing the development of modules with greater efficiency and lower cost.

Thermoelectric technologies are essential for numerous low-power applications, such as wearable electronics, autonomous sensors for environment monitoring, devices for medical monitoring, diagnostics and treatment, and the IoT. They can contribute to energy efficiency through waste-heat recovery into electricity and enable the supply of power in remote areas. Thus, thermoelectric technologies support and enable environmentally friendly and sustainable development for a better quality of life. It is difficult to imagine a future without the benefits that thermoelectric technologies could bring. So, can we afford not to invest in the further development of thermoelectric technologies?

7.5. References

Abas, N., Kalair, A., Khan, N. (2015). Review of fossil fuels and future energytechnologies. *Futures*, 69, 31–49.

Ahmad, T. and Zhang, D. (2020). A critical review of comparative global historical energy consumption and future demand: The story told so far. *Energy Reports*, 6, 1973–1991.

Akinaga, H. (2020). Recent advances and future prospects in energy harvesting technologies. *Japanese Journal of Applied Physics*, 59, 110201.

Amollo, T.A., Mola, G.T., Kirui, M.S.K., Nyamori, V.O. (2017). Graphene for thermoelectric applications: Prospects and challenges. *Critical Reviews in Solid State and Materials Sciences*, 43, 133–157.

Aswal, D.K., Basu, R., Singh, A. (2016). Key issues in development of thermoelectric power generators: High figure-of-merit materials and their highly conducting interfaces with metallic interconnects. *Energy Conversion and Management*, 114, 50–67.

Atta, R.M. (2018). Thermoelectric cooling. In *Bringing Thermoelectricity Into Reality*, Patricia Aranguren, P. (ed.). IntechOpen, London.

Bahk, J.-H., Fang, H., Yazawa, K., Shakouri, A. (2015). Flexible thermoelectric materials and device optimization for wearable energy harvesting. *Journal of Materials Chemistry C*, 3, 10362–10374.

Bai, Y., Jantunen, H., Juuti, J. (2018). Energy harvesting research: The road from single source to multisource. *Advanced Materials*, 1707271.

Bai, S., Zhang, Q., Chen, L. (2021). Segmented modules. In *Thermoelectric Energy Conversion*, Funahashi, R. (ed.). Elsevier, Amsterdam.

Baláž, P., Baláž, M., Achimovičová, M., Bujňáková, Z., Dutková, E. (2017). Chalcogenide mechanochemistry in materials science: Insight into synthesis and applications (a review). *Journal of Materials Science*, 52, 11851–11890.

Baláž, P., Hegedüs, M., Reece, M., Zhang, R.-Z., Su, T., Škorvánek, I., Briančin, J., Baláž, M., Mihálik, M., Tešínsky, M. et al. (2019). Mechanochemistry for thermoelectrics: Nanobulk $Cu_6Fe_2SnS_8/Cu_2FeSnS_4$ composite synthesized in an industrial mill. *Journal of Electronic Materials*, 48, 1846–1856.

Baláž, P., Guilmeau, E., Daneu, N., Dobrozhan, O., Baláž, M., Hegedus, M., Barbier, T., Achimovičová, M., Kaňuchová, M., Briančin, J. (2020). Tetrahedrites synthesized via scalable mechanochemical process and spark plasma sintering. *Journal of the European Ceramic Society*, 40, 1922–1930.

Baru, S. and Bhatia, S. (2020). A review on thermoelectric cooling technology and its applications. In *IOP Conference Series: Materials Science and Engineering*, 912, 042004.

Bernik, S. (2021). Oxide thermoelectric materials: Compositional, structural, microstructural, and processing challanges to realize their potential. In *Thermoelectric Energy Conversion Theories and Mechanisms, Materials, Devices, and Applications*, Funahashi, R. (ed.). Elsevier, Amsterdam.

Blackburn, J.L., Ferguson, A.J., Cho, C., Grunlan, J.C. (2018). Carbon-nanotube-based thermoelectric materials and devices. *Advanced Materials*, 30, 1704386.

Boukai, A.I., Bunimovich, Y., Tahir-Kheli, J., Yu, J.K., Goddard III, W.A., Heath, J.R. (2008). Silicon nanowires as efficient thermoelectric materials. *Nature*, 451, 168–171.

Burkov, A.T. (2018). Silicide thermoelectrics: Materials for energy harvesting. *Physica Status Solidi (A)*, 215, 1800105.

Bux, S.K., Yeung, M.T., Toberer, E.S., Snyder, G.J., Kaner, R.B., Fleurial, J.-P. (2011). Mechanochemical synthesis and thermoelectric properties of high quality magnesium silicide. *Journal of Materials Chemistry*, 21, 12259–12266.

Cai, B., Hu, H., Zhuang, H.-L., Li, J.-F. (2019). Promising materials for thermoelectric applications. *Journal of Alloys and Compounds*, 806, 471–486.

Chakraborty, P., Ma, T., Zahiri, A.H., Cao, L., Wang, Y. (2018). Carbon-based materials for thermoelectrics. *Advances in Condensed Matter Physics*, 2018, 3898479, 1–29.

Champier, D. (2017). Thermoelectric generators: A review of applications. *Energy Conversion and Management*, 140, 167–181.

Charris, D., Gomez, D., Ortega, A.R., Carmona, M., Pardo, M. (2020). A thermoelectric energy harvesting scheme with passive cooling for outdoor IoT sensors. *Energies*, 13, 2782.

Chen, S. and Ren, Z. (2013). Recent progress of half-Heusler for moderate temperature thermoelectric applications. *Materials Today*, 16, 387–395.

Chen, Z.-G., Han, G., Yang, L., Cheng, L., Zou, J. (2012). Nanostructured thermoelectric materials: Current research and future challenge. *Progress in Natural Science: Materials International*, 22, 535–549.

Chen, K.-X., Li, M.-S., Mo, D.-C., Lyu, S.-S. (2018). Nanostructural thermoelectric materials and their performance. *Frontiers in Energy*, 12, 97–108.

Chetty, R., Bali, A., Mallik, R.C. (2015). Tetrahedrites as thermoelectric materials: An overview. *Journal of Materials Chemistry C*, 3, 12364–12378.

Christensen, M., Johnsen, S., Iversen, B.B. (2010). Thermoelectric clathrates of type I. *Dalton Trans.*, 39, 978–992.

Ciesielski, K., Synoradzki, K., Veremchuk, I., Skokowski, P., Szymański, D., Grin, Y., Kaczorowski, D. (2020). Thermoelectric performance of the half-Heusler phases RNiSb (R=Sc,Dy,Er,Tm,Lu): High mobility ratio between majority and minority charge carriers. *Physical Review Applied*, 14, 054046.

Dongxu, J., Zhongbao, W., Pou, J., Mazzoni, S., Rajoo, S., Romagnoli, A. (2019). Geometry optimization of thermoelectric modules: Simulation and experimental study. *Energy Conversion and Management*, 195, 236–243.

El-Desouky, A., Carter, M., Andre, M.A., Bardet, P.M., Leblanc, S. (2016). Rapid processing and assembly of semiconductor thermoelectric materials for energy conversion devices. *Materials Letters*, 185, 598–602.

Ellabban, O., Abu-Rub, H., Blaabjerg, F. (2014). Renewable energy resources: Current status, future prospects and their enabling technology. *Renewable and Sustainable Energy Reviews*, 39, 748–764.

Feng, Y., Jiang, X., Ghafari, E., Kucukgok, B., Zhang, C., Ferguson, I., Lu, N. (2017). Metal oxides for thermoelectric power generation and beyond. *Advanced Composites and Hybrid Materials*, 1, 114–126.

Fernández-Yáñez, P., Romero, V., Armas, O., Cerretti, G. (2021). Thermal management of thermoelectric generators for waste energy recovery. *Applied Thermal Engineering*, 196, 117291.

Ferreira-Teixeira, S. and Pereira, A.M. (2018). Geometrical optimization of a thermoelectric device: Numerical simulations. *Energy Conversion and Management*, 169, 217–227.

Freer, R. and Powell, A.V. (2020). Realising the potential of thermoelectric technology: A roadmap. *Journal of Materials Chemistry C*, 8, 441–463.

Gao, H.B., Huang, G.H., Li, H.J., Qu, Z.G., Zhang, Y.J. (2016). Development of stove-powered thermoelectric generators: A review. *Applied Thermal Engineering*, 96, 297–310.

Gould, C. (2019). Thermoelectric energy harvesting. In *A Guide to Small-Scale Energy Harvesting Techniques*, Manyala, R. (ed.). IntechOpen, London.

Guélou, G., Lemoine, P., Raveau, B., Guilmeau, E. (2021). Recent developments in high-performance thermoelectric sulphides: An overview of the promising synthetic colusites. *Journal of Materials Chemistry C*, 9, 773–795.

Gurevich, Y.G. and Velazquez-Perez, J.E. (2014). Peltier effect in semiconductors. In *Wiley Encyclopedia of Electrical and Electronics Engineering*, Webster, J.G. (ed.). John Wiley & Sons, Inc, New York.

Guttmann, G.M. and Gelbstein, Y. (2018). Mechanical properties of thermoelectric materials for practical applications. In *Bringing Thermoelectricity Into Reality*, Aranguren, P. (ed.). IntechOpen, London.

Hamawandi, B., Mansouri, H., Ballikaya, S., Demirci, Y., Orlovská, M., Bolghanabadi, N., Sajjadi, S.A., Toprak, M.S. (2020). A comparative study on the thermoelectric properties of bismuth chalcogenide alloys synthesized through mechanochemical alloying and microwave-assisted solution synthesis routes. *Frontiers in Materials*, 7, 569723.

Han, C., Li, Z., Dou, S. (2014). Recent progress in thermoelectric materials. *Chinese Science Bulletin*, 59, 2073–2091.

He, J., Kanatzidis, M.G., Dravid, V.P. (2013). High performance bulk thermoelectrics via a panoscopic approach. *Materials Today*, 16, 166–176.

He, R., Gahlawat, S., Guo, C., Chen, S., Dahal, T., Zhang, H., Liu, W., Zhang, Q., Chere, E., White, K. et al. (2015). Studies on mechanical properties of thermoelectric materials by nanoindentation. *Physica Status Solidi (A)*, 212, 2191–2195.

He, R., Heyn, W., Thiel, F., Pérez, N., Damm, C., Pohl, D., Rellinghaus, B., Reimann, C., Beier, M., Friedrich, J. et al. (2019). Thermoelectric properties of silicon and recycled silicon sawing waste. *Journal of Materiomics*, 5, 15–33.

Hochbaum, A.I., Chen, R., Delgado, R.D., Liang, W., Garnett, E.C., Najarian, M., Majumdar, A., Yang, P. (2008). Enhanced thermoelectric performance of rough silicon nanowires. *Nature*, 451, 163–167.

Hung, N.T., Nugraha, A.R.T., Saito, R. (2019). Thermoelectric properties of carbon nanotubes. *Energies*, 12, 4561.

Jaldurgam, F.F., Ahmad, Z., Touati, F. (2021). Synthesis and performance of large-scale cost-effective environment-friendly nanostructured thermoelectric materials. *Nanomaterials (Basel)*, 11, 1091.

Jaziri, N., Boughamoura, A., Müller, J., Mezghani, B., Tounsi, F., Ismail, M. (2020). A comprehensive review of thermoelectric generators: Technologies and common applications. *Energy Reports*, 6, 264–287.

Ji, L. (2018). Metal oxide-based thermoelectric materials. In *Metal Oxides in Energy Technologies*, Wu, Y. (ed.). Elsevier, Amsterdam.

Karana, D.R. and Sahoo, R.R. (2019). Influence of geometric parameter on the performance of a new asymmetrical and segmented thermoelectric generator. *Energy*, 179, 90–99.

Kenfaui, D., Chateigner, D., Gomina, M., Noudem, J.G. (2010). Texture, mechanical and thermoelectric properties of $Ca_3Co_4O_9$ ceramics. *Journal of Alloys and Compounds*, 490, 472–479.

Kim, H.S., Wang, T., Liu, W., Ren, Z. (2016). Engineering thermal conductivity for balancing between reliability and performance of bulk thermoelectric generators. *Advanced Functional Materials*, 26, 3678–3686.

Kim, G., Kim, W., Lee, W. (2019). Fabrication of silicide-based thermoelectric nanocomposites: A review. *Journal of the Korean Ceramic Society*, 56, 435–442.

Kim, G., Shin, H., Lee, J., Lee, W. (2021). A review on silicide-based materials: Thermoelectric and mechanical properties. *Metals and Materials International*, 27, 2205–2219.

Kishore, R.A. and Priya, S. (2018). A review on low-grade thermal energy harvesting: Materials, methods and devices. *Materials (Basel)*, 11, 1433.

Kruszewski, M.J., Ciupiński, Ł., Zybała, R. (2021). Review of rapid fabrication methods of skutterudite materials. *Materials Today: Proceedings*, 44, 3475–3482.

Kumar, P.M., Babu, V.J., Subramanian, A., Bandla, A., Thakor, N., Ramakrishna, S., Wei, H. (2019). The design of a thermoelectric generator and its medical applications. *Designs*, 3, 22.

Leblanc, S. (2014). Thermoelectric generators: Linking material properties and systems engineering for waste heat recovery applications. *Sustainable Materials and Technologies*, 1–2, 26–35.

Leblanc, S., Yee, S.K., Scullin, M.L., Dames, C., Goodson, K.E. (2014). Material and manufacturing cost considerations for thermoelectrics. *Renewable and Sustainable Energy Reviews*, 32, 313–327.

Lee, H. (2013). Optimal design of thermoelectric devices with dimensional analysis. *Applied Energy*, 106, 79–88.

Lei, Y., Gao, W., Zheng, R., Li, Y., Chen, W., Ma, L., Wan, R., Zhou, H. (2019). Microstructure and thermoelectric performance of n-type $Co_4Sb_{11.3}Te_{0.5}Se_{0.2}$ skutterudite synthesized via microwave heating. *AIP Advances*, 9, 025124.

Li, J.-F., Liu, W.-S., Zhao, L.-D., Zhou, M. (2010). High-performance nanostructured thermoelectric materials. *NPG Asia Materials*, 2, 152–158.

Li, J., Pan, Y., Wu, C., Sun, F., Wei, T. (2017). Processing of advanced thermoelectric materials. *Science China Technological Sciences*, 60, 1347–1364.

Li, S., Li, X., Ren, Z., Zhang, Q. (2018). Recent progress towards high performance of tin chalcogenide thermoelectric materials. *Journal of Materials Chemistry A*, 6, 2432–2448.

Li, C., Jiang, F., Liu, C., Liu, P., Xu, J. (2019). Present and future thermoelectric materials toward wearable energy harvesting. *Applied Materials Today*, 15, 543–557.

Li, G., An, Q., Duan, B., Borgsmiller, L., Malki, M.A., Agne, M., Aydemir, U., Zhai, P., Zhang, Q., Morozov, S.I. et al. (2021). Fracture toughness of thermoelectric materials. *Materials Science and Engineering: R: Reports*, 144, 100607.

Lindorf, M., Mazzio, K.A., Pflaum, J., Nielsch, K., Brütting, W., Albrecht, M. (2020). Organic-based thermoelectrics. *Journal of Materials Chemistry A*, 8, 7495–7507.

Liu, W. and Bai, S. (2019). Thermoelectric interface materials: A perspective to the challenge of thermoelectric power generation module. *Journal of Materiomics*, 5, 321–336.

Liu, W., Jie, Q., Kim, H.S., Ren, Z. (2015). Current progress and future challenges in thermoelectric power generation: From materials to devices. *Acta Materialia*, 87, 357–376.

Liu, W., Yin, K., Zhang, Q., Uher, C., Tang, X. (2017). Eco-friendly high-performance silicide thermoelectric materials. *National Science Review*, 4, 611–626.

Liu, Z.-Y., Zhu, J.-L., Tong, X., Niu, S., Zhao, W.-Y. (2020). A review of CoSb3-based skutterudite thermoelectric materials. *Journal of Advanced Ceramics*, 9, 647–673.

Liu, H., Fu, H., Sun, L., Lee, C., Yeatman, E.M. (2021). Hybrid energy harvesting technology: From materials, structural design, system integration to applications. *Renewable and Sustainable Energy Reviews*, 137, 110473.

Lorenzi, B., Mariani, P., Reale, A., Di Carlo, A., Chen, G., Narducci, D. (2021). Practical development of efficient thermoelectric – Photovoltaic hybrid systems based on wide-gap solar cells. *Applied Energy*, 300, 117343.

Ma, Q., Fang, H., Zhang, M. (2017). Theoretical analysis and design optimization of thermoelectric generator. *Applied Thermal Engineering*, 127, 758–764.

Ma, Z., Wei, J., Song, P., Zhang, M., Yang, L., Ma, J., Liu, W., Yang, F., Wang, X. (2021). Review of experimental approaches for improving zT of thermoelectric materials. *Materials Science in Semiconductor Processing*, 121, 105303.

Masood, K.B., Kumar, P., Singh, R.A., Singh, J. (2018). Odyssey of thermoelectric materials: Foundation of the complex structure. *Journal of Physics Communications*, 2, 062001.

Meng, J.-H., Wang, X.-D., Chen, W.-H. (2016). Performance investigation and design optimization of a thermoelectric generator applied in automobile exhaust waste heat recovery. *Energy Conversion and Management*, 120, 71–80.

Mohanraman, R., Lan, T.W., Hsiung, T.C., Amada, D., Lee, P.C., Ou, M.N., Chen, Y.Y. (2015). Engineering nanostructural routes for enhancing thermoelectric performance: Bulk to nanoscale. *Front. Chem.*, 3, 63.

Ni, J.E., Case, E.D., Schmidt, R.D., Wu, C.-I., Hogan, T.P., Trejo, R.M., Kirkham, M.J., Lara-Curzio, E., Kanatzidis, M.G. (2013). The thermal expansion coefficient as a key design parameter for thermoelectric materials and its relationship to processing-dependent bloating. *Journal of Materials Science*, 48, 6233–6244.

Novitskii, A., Serhiienko, I., Nepapushev, A., Ivanova, A., Sviridova, T., Moskovskikh, D., Voronin, A., Miki, H., Khovaylo, V. (2021). Mechanochemical synthesis and thermoelectric properties of $TiFe_2Sn$ Heusler alloy. *Intermetallics*, 133, 107195.

Nozariasbmarz, A., Agarwal, A., Coutant, Z.A., Hall, M.J., Liu, J., Liu, R., Malhotra, A., Norouzzadeh, P., Öztürk, M.C., Ramesh, V.P. et al. (2017). Thermoelectric silicides: A review. *Japanese Journal of Applied Physics*, 56. 05DA04.

Nozariasbmarz, A., Kishore, R.A., Poudel, B., Saparamadu, U., Li, W., Cruz, R., Priya, S. (2019). High power density body heat energy harvesting. *ACS Applied Materials & Interfaces*, 11, 40107–40113.

Nozariasbmarz, A., Collins, H., Dsouza, K., Polash, M.H., Hosseini, M., Hyland, M., Liu, J., Malhotra, A., Ortiz, F.M., Mohaddes, F. et al. (2020a). Review of wearable thermoelectric energy harvesting: From body temperature to electronic systems. *Applied Energy*, 258, 114069.

Nozariasbmarz, A., Suarez, F., Dycus, J.H., Cabral, M.J., Lebeau, J.M., Öztürk, M.C., Vashaee, D. (2020b). Thermoelectric generators for wearable body heat harvesting: Material and device concurrent optimization. *Nano Energy*, 67, 104265.

Orrill, M. and Leblanc, S. (2017). Printed thermoelectric materials and devices: Fabrication techniques, advantages, and challenges. *Journal of Applied Polymer Science*, 134, 44256.

Paravantis, J.A. and Kontouli, N. (2020). Energy security and renewable energy: A geopolitical perspective. In *Renewable Energy Resources, Challenges and Applications*, Al Qubeissi, M. (ed.). IntechOpen, London.

Peng, S., Wang, D., Lu, J., He, M., Xu, C., Li, Y., Zhu, S. (2017). A review on organic polymer-based thermoelectric materials. *Journal of Polymers and the Environment*, 25, 1208–1218.

Pham, N.H., Vallin, Ö., Panda, J., Venkata Kamalakar, M., Guo, J., Luo, J., Wen, C., Zhang, S.-L., Zhang, Z.-B. (2020). High thermoelectric power factor of p-type amorphous silicon thin films dispersed with ultrafine silicon nanocrystals. *Journal of Applied Physics*, 127, 245304.

Poon, S. (2018). Recent advances in thermoelectric performance of half-Heusler compounds. *Metals*, 8, 989.

Powell, A.V. (2019). Recent developments in Earth-abundant copper-sulfide thermoelectric materials. *Journal of Applied Physics*, 126, 100901.

Prauzek, M., Konecny, J., Borova, M., Janosova, K., Hlavica, J., Musilek, P. (2018). Energy harvesting sources, storage devices and system topologies for environmental wireless sensor networks: A review. *Sensors (Basel)*, 18, 2446.

Qin, Y., Yang, L., Wei, J., Yang, S., Zhang, M., Wang, X., Yang, F. (2020). Doping effect on Cu_2Se thermoelectric performance: A review. *Materials (Basel)*, 13, 5704.

Ravi, V., Firdosy, S., Caillat, T., Brandon, E., Van Der Walde, K., Maricic, L., Sayir, A. (2009). Thermal expansion studies of selected high-temperature thermoelectric materials. *Journal of Electronic Materials*, 38, 1433–1442.

Recatala-Gomez, J., Suwardi, A., Nandhakumar, I., Abutaha, A., Hippalgaonkar, K. (2020). Toward accelerated thermoelectric materials and process discovery. *ACS Applied Energy Materials*, 3, 2240–2257.

Rogl, G. and Rogl, P. (2017). Skutterudites, a most promising group of thermoelectric materials. *Current Opinion in Green and Sustainable Chemistry*, 4, 50–57.

Rogl, G., Grytsiv, A., Rogl, P., Bauer, E., Hochenhofer, M., Anbalagan, R., Mallik, R.C., Schafler, E. (2014). Nanostructuring of p- and n-type skutterudites reaching figures of merit of approximately 1.3 and 1.6, respectively. *Acta Materialia*, 76, 434–448.

Rojas, J.P., Singh, D., Inayat, S.B., Sevilla, G.A.T., Fahad, H.M., Hussain, M.M. (2017). Review–micro and nano-engineering enabled new generation of thermoelectric generator devices and applications. *ECS Journal of Solid State Science and Technology*, 6, N3036–N3044.

Rowe, D.M. (1995). *CRC Handbook of Thermoelectrics*. CRC Press, Boca Raton.

Rull-Bravo, M., Moure, A., Fernández, J.F., Martín-González, M. (2015). Skutterudites as thermoelectric materials: Revisited. *RSC Advances*, 5, 41653–41667.

Russ, B., Glaudell, A., Urban, J.J., Chabinyc, M.L., Segalman, R.A. (2016). Organic thermoelectric materials for energy harvesting and temperature control. *Nature Reviews Materials*, 1, 1–14.

Ryu, H., Yoon, H.J., Kim, S.W. (2019). Hybrid energy harvesters: Toward sustainable energy harvesting. *Advanced Materials*, 31, 1802898.

Salah, W.A. and Abuhelwa, M. (2020). Review of thermoelectric cooling devices recent applications. *Journal of Engineering Science and Technology*, 15, 455–476.

Salvardi, M.S. and Salvardi, H. (2020). For sustainable development: Future trends in renewable energy and enabling technologies. In *Renewable Energy Resources, Challenges and Applications*, Al Qubeissi, M. (ed.). IntechOpen, London.

Schierning, G., Chavez, R., Schmechel, R., Balke, B., Rogl, G., Rogl, P. (2015). Concepts for medium-high to high temperature thermoelectric heat-to-electricity conversion: A review of selected materials and basic considerations of module design. *Translational Materials Research*, 2, 025001.

Shi, Y., Sturm, C., Kleinke, H. (2019). Chalcogenides as thermoelectric materials. *Journal of Solid State Chemistry*, 270, 273–279.

Shi, X.L., Zou, J., Chen, Z.G. (2020). Advanced thermoelectric design: From materials and structures to devices. *Chemical Reviews*, 120, 7399–7515.

Shtern, Y.I., Rogachev, M.S., Bublik, V.T., Tarasova, I.V., Pozdniakov, A.V. (2019). The results of thermal expansion investigation for effective TE materials. *IEEE Xplore*, 1932–1936.

Siddique, A.R.M., Mahmud, S., Heyst, B.V. (2017). A review of the state of the science on wearable thermoelectric power generators (TEGs) and their existing challenges. *Renewable and Sustainable Energy Reviews*, 73, 730–744.

Snyder, J.G. and Toberer, S. (2008). Complex thermoelectric materials. *Nature Materials*, 7, 105–114.

Snyder, G.J. and Ursell, T.S. (2003). Thermoelectric efficiency and compatibility. *Physical Review Letters*, 91, 148301.

Stančin, H., Mikulčić, H., Wang, X., Duić, N. (2020). A review on alternative fuels in future energy system. *Renewable and Sustainable Energy Reviews*, 128, 109927.

Sui, F. and Kauzlarich, S.M. (2016). Tuning thermoelectric properties of type I clathrate $K_{8-x}Ba_xA_{18+x}Si_{38-x}$ through barium substitution. *Chemistry of Materials*, 28, 3099–3107.

Takabatake, T., Suekuni, K., Nakayama, T., Kaneshita, E. (2014). Phonon-glass electron-crystal thermoelectric clathrates: Experiments and theory. *Reviews of Modern Physics*, 86, 669–716.

Tang, J., Wang, H.T., Lee, D.H., Fardy, M., Huo, Z., Russell, T.P., Yang, P. (2010). Holey silicon as an efficient thermoelectric material. *Nano Lett.*, 10, 4279–4283.

Tang, X., Wang, X., Cattley, R., Gu, F., Ball, A.D. (2018). Energy harvesting technologies for achieving self-powered wireless sensor networks in machine condition monitoring: A review. *Sensors (Basel)*, 18, 4113.

Tanusilp, S.A. and Kurosaki, K. (2019). Si-based materials for thermoelectric applications. *Materials (Basel)*, 12, 1943.

Tanwar, A., Lal, S., Razeeb, K. (2021). Structural design optimization of micro-thermoelectric generator for wearable biomedical devices. *Energies*, 14, 2339.

Thimont, Y. and Leblanc, S. (2019). The impact of thermoelectric leg geometries on thermal resistance and power output. *Journal of Applied Physics*, 126, 095101.

Wang, B.L. and Li, J. (2020). Fracture behavior and thermal shock resistance analysis of thermoelectric material plates and shells under thermal and electric shocks. *Engineering Fracture Mechanics*, 225, 106130.

Wang, H. and Yu, C. (2019). Organic thermoelectrics: Materials preparation, performance optimization, and device integration. *Joule*, 3, 53–80.

Wang, B.L., Guo, Y.B., Zhang, C.W. (2016). Cracking and thermal shock resistance of a Bi_2Te_3 based thermoelectric material. *Engineering Fracture Mechanics*, 152, 1–9.

Wang, X., Wang, H., Liu, B. (2018). Carbon nanotube-based organic thermoelectric materials for energy harvesting. *Polymers (Basel)*, 10, 1196.

Wei, T.-R., Wu, C.-F., Li, F., Li, J.-F. (2018). Low-cost and environmentally benign selenides as promising thermoelectric materials. *Journal of Materiomics*, 4, 304–320.

Wereszczak, A.A. and Case, E.D. (2015). Mechanical response of thermoelectric materials. Report, Oak Ridge National Laboratory, Oak Ridge.

Wolf, M., Rybakov, A., Hinterding, R., Feldhoff, A. (2020). Geometry optimization of thermoelectric modules: Deviation of optimum power output and conversion efficiency. *Entropy (Basel)*, 22, 1233.

Wu, S., Jiang, J., Liang, Y., Yang, P., Niu, Y., Chen, Y., Xia, J., Wang, C. (2017). Chemical precipitation synthesis and thermoelectric properties of copper sulfide. *Journal of Electronic Materials*, 46, 2432–2437.

Yan, J., Liao, X., Yan, D., Chen, Y. (2018). Review of micro thermoelectric generator. *Journal of Microelectromechanical Systems*, 27, 1–18.

Yao, H., Fan, Z., Cheng, H., Guan, X., Wang, C., Sun, K., Ouyang, J. (2018). Recent development of thermoelectric polymers and composites. *Macromol. Rapid Commun.*, 39, 1700727.

Yao, C.J., Zhang, H.L., Zhang, Q. (2019). Recent progress in thermoelectric materials based on conjugated polymers. *Polymers (Basel)*, 11, 107.

Young, J.S. and Reddy, R.G. (2019). Processing and thermoelectric properties of TiNiSn materials: A review. *Journal of Materials Engineering and Performance*, 28, 5917–5930.

Yu, K., Zhou, Y., Liu, Y., Liu, F., Hu, L., Ao, W., Zhang, C., Li, Y., Li, J., Xie, H. (2019a). Near-room-temperature thermoelectric materials and their application prospects in geothermal power generation. *Geomechanics and Geophysics for Geo-Energy and Geo-Resources*, 6, 12.

Yu, Y., Zhu, W., Kong, X., Wang, Y., Zhu, P., Deng, Y. (2019b). Recent development and application of thin-film thermoelectric cooler. *Frontiers of Chemical Science and Engineering*, 14, 492–503.

Yu, Y., Cagnoni, M., Cojocaru-Mirédin, O., Wuttig, M. (2020). Chalcogenide thermoelectrics empowered by an unconventional bonding mechanism. *Advanced Functional Materials*, 30, 1904862.

Zabek, D. and Morini, F. (2019). Solid state generators and energy harvesters for waste heat recovery and thermal energy harvesting. *Thermal Science and Engineering Progress*, 9, 235–247.

Zhang, Y. and Park, S.J. (2019). Flexible organic thermoelectric materials and devices for wearable green energy harvesting. *Polymers (Basel)*, 11, 909.

Zhang, Q., Sun, Y., Xu, W., Zhu, D. (2014). Organic thermoelectric materials: Emerging green energy materials converting heat to electricity directly and efficiently. *Advanced Materials*, 26, 6829–6851.

Zhang, Y., Heo, Y.J., Park, M., Park, S.J. (2019). Recent advances in organic thermoelectric materials: Principle mechanisms and emerging carbon-based green energy materials. *Polymers (Basel)*, 11, 167.

Zhao, D. and Tan, G. (2014). A review of thermoelectric cooling: Materials, modeling and applications. *Applied Thermal Engineering*, 66, 15–24.

Zhu, K., Deng, B., Zhang, P., Kim, H.S., Jiang, P., Liu, W. (2020). System efficiency and power: The bridge between the device and system of a thermoelectric power generator. *Energy & Environmental Science*, 13, 3514–3526.

Zong, P.-A., Liang, J., Zhang, P., Wan, C., Wang, Y., Koumoto, K. (2020). Graphene-based thermoelectrics. *ACS Applied Energy Materials*, 3, 2224–2239.

Zoui, M.A., Bentouba, S., Stocholm, J.G., Bourouis, M. (2020). A review on thermoelectric generators: Progress and applications. *Energies*, 13, 3606.

8

Processing of Thermoelectric Transition Metal Silicides Towards Module Development

Sylvain LE TONQUESSE[1], Mathieu PASTUREL[2], Franck GASCOIN[3] and David BERTHEBAUD[1]

[1] *CNRS-Saint Gobain-NIMS, IRL 3629, Laboratory for Innovative Key Materials and Structures (LINK), National Institute for Materials Science (NIMS), Tsukuba, Japan*

[2] *Institut des Sciences Chimiques de Rennes, UMR CNRS 6226, Université de Rennes, France*

[3] *ENSICAEN, UNICAEN, CNRS, CRISMAT, Normandie Université, Caen, France*

8.1. Introduction

Among thermoelectric materials, the group of transition metal silicides is neither the most studied nor the most performing. Nevertheless, in the last decade, they have been attracting for researchers from both academic and industry as they possess many interesting fundamental features and advantages which we will develop in the upcoming chapter. So far, the most commonly studied transition metals silicide materials (ex: $CrSi_2$, β-$FeSi_2$, $MnSi_\gamma$ with $\gamma \sim 1.74$) have been chosen for their potential as bulk materials since such compounds are not made of critical metals, are low cost,

For a color version of all the figures in this chapter, see www.iste.co.uk/akinaga/thermoelectric2.zip.

Thermoelectric Micro/Nano Generators 2,
coordinated by Hiroyuki AKINAGA, Atsuko KOSUGA,
Takao MORI and Gustavo ARDILA.

environmentally friendly, suitable materials for mid-range temperature applications, and for some of them, possess a relatively high average figure of merit zT over a large temperature range. Despite those advantages, there is still a need to improve their thermoelectric performances as well as to develop new process which would be suitable for the industrial process and mass production. In this chapter, we will present recent advances on the processing of transition metal silicides for thermoelectricity through various techniques, such as magnesiothermic reduction (MR) synthesis, mechanical alloying, melting processes and additive manufacturing. The above-mentioned techniques not only enables the improvement of thermoelectric performance through various effects such as nano(meso)structuration or preferential orientation, but can also be promising processes for large-scale production of materials, which is an essential criterion to determine industrializable candidates for future applications. We will also emphasize the importance of a better understanding of the crystal structure of these materials and its evolution during operating conditions through some examples of advanced structural characterization studies of transition metal silicides. Finally, we will shortly present recent studies on module/leg development of silicide materials which will hopefully pave the way for future applications.

8.2. Recent progress on the process of thermoelectric transition metals silicide

8.2.1. *Synthesis of mesostructured silicides through magnesiothermic reduction*

Owing to its simplicity and low-temperature conditions, MR is a less energy-intensive synthesis method with a strong potential for industrial scale. Processes of MR consist of the co-reduction of metal oxides by magnesium in order to form alloys or intermetallics according to the reaction:

$$\mathbf{MO_x + M'O^y + (X + y)Mg \xrightarrow{\Delta} MM' + (x + y)MgO} \qquad [8.1]$$

This type of reaction has been exploited on an industrial scale (Kroll or Ames processes) but very little explored for the synthesis of thermoelectric intermetallics, since it has many advantages over conventional fusion/solidification synthesis methods: it requires oxide precursors generally cheaper and more stable in air than their metallic counterpart, lower synthesis temperatures and enables us to obtain submicron powders suitable for the densification of mesostructured materials. Recently, this method has been applied to synthesize both pure and doped thermoelectric silicides – β-$FeSi_2$ and $MnSi_\gamma$ (Le Tonquesse et al. 2019, 2020a). Indeed, although these materials are good candidates for large-scale industrialization, their synthesis by fusion/solidification is very energy-intensive because of their peritectic or peritectoid formations and the improvement of their zT necessarily requires a consequent reduction of their high thermal conductivities through nano- or mesostructuration.

The reliable and reproducible synthesis procedure consists of placing pelletized oxide precursors with magnesium in a molybdenum crucible in order to avoid any reactivity with the reactor. The grain size and morphologies of the precursor oxide powders playing an essential role in the success of the reaction, it is first necessary to fully optimize the synthesis step parameters through reaction mechanism analysis with the help of X-ray diffraction study. After this first optimization step, the fast heat treatment (less than 24 h including heating ramp, dwell and cooling) performed below the decomposition temperature offers the possibility to produce undoped and doped compounds with controlled compositions (e.g. $V_{0.04}Mn_{0.96}Si_{1.74}$ and $Co_{0.07}Fe_{0.93}Si_2$) (Le Tonquesse et al. 2019, 2020a). Finally, the obtained powders are made of very small (from 20 to 400 nm) and well-crystallized particles which are suitable for the shaping of mesostructured spark plasma sintered materials with an average grain size of approximately 500 nm. The obtained mesostructured compounds were shown to reliably exhibit a reduction of the lattice thermal conductivity by up to 20% for $MnSi_{1.74}$ and β-$FeSi_2$ compared to conventionally arc-melted silicides. However, this beneficial effect is counterbalanced by a decrease of the power factor. Nevertheless, for $MnSi_{1.74}$, it eventually results in a similar $zT_{max} = 0.4$ at 750 K, while for β-$Co_{0.07}Fe_{0.93}Si_2$, zT is improved with a maximum value reaching 0.18 at 773 K in comparison to arc-melted compounds.

To gain better understanding of the mesostructured effect on the thermoelectric properties, advanced structural characterization of these silicide materials has been carried out. Indeed, these materials possess a complex crystal structure which requires in-depth crystallographic study to grasp their structure–properties relationship. In the case of $MnSi_\gamma$ ($1.725 < \gamma < 1.75$), also known as higher manganese silicide (HMS), it crystallizes in a Nowotny chimney-ladder structure-type composite composed of two tetragonal subsystems of Mn and Si generating an incommensurate modulated structure (Miyazaki et al. 2008). This modulated structure can be described by a modulation vector q = (00γ) and a space group in 3 + 1 dimensions. It was shown that the γ component of the modulation vector corresponds exactly to the overall stoichiometry of the compound. This one could be precisely determined by a Rietveld refinement from powder XRD diagrams (Figure 8.1). Such X-ray characterization highlights the homogeneity of the materials obtained through magnesiothermy synthesis which prevents the formation of the MnSi by-product usually found as a secondary phase after solidification of the melt. This metallic secondary phase grows perpendicularly to the *c*-axis of $MnSi_γ$ and is known to be detrimental to the thermoelectric performance of the material, degrading its thermopower while increasing its thermal conductivity. Therefore, the absence of such secondary phase plays a great role in the remarkable decrease of thermal conductivity between MR and arc-melted MnSiγ (Le Tonquesse et al. 2020a).

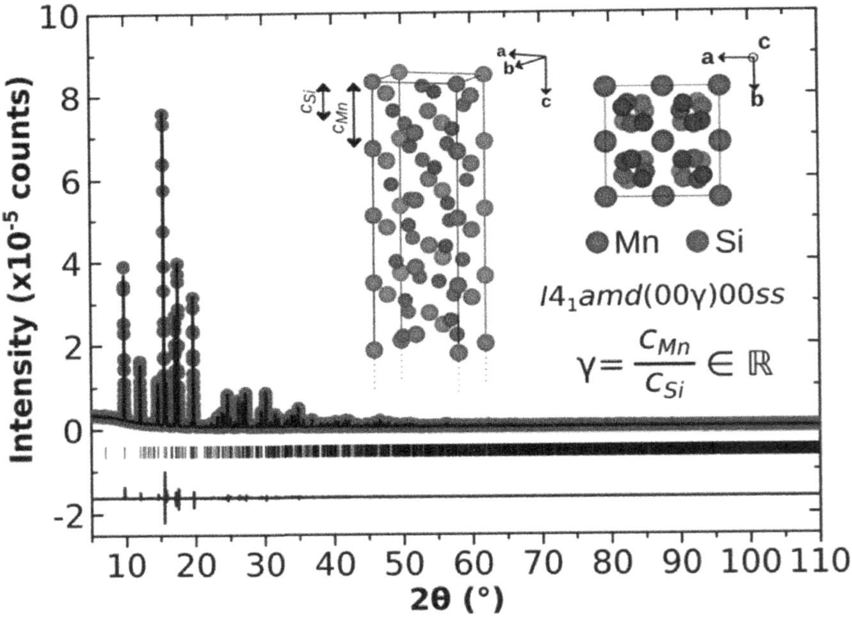

Figure 8.1. *X-ray synchrotron diffractogram of $MnSi_γ$ refined by the Rietveld method with the JANA2006 software (Petricek et al. 2014). The inset illustrates the composite structure of MnSiγ. From Le Tonquesse et al. (2020a)*

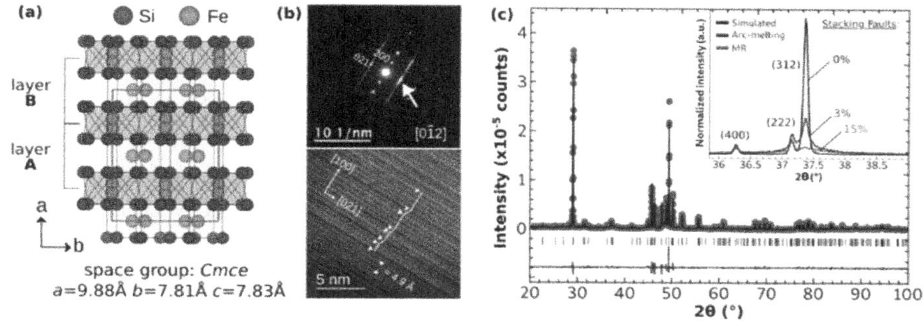

Figure 8.2. *(a) β-$FeSi_2$ crystal structure. (b) ED and HRTEM images of a β-$FeSi_2$ grain highlighting the presence of stacking faults. (c) Rietveld refinements of β-$FeSi_2$ X-ray diffraction patterns with the help of the FAULTS software (Casas-Cabanas et al. 2016) implemented in the FullProf Suite (Rodriguez-Carvajal 1993). From Le Tonquesse et al. (2019)*

As for $MnSi_\gamma$, the notable effect on the thermal conductivity was observed for β-$FeSi_2$ and therefore an in-depth crystal structure study was carried out to assess the structure/properties relationship. Contrary to $MnSi_\gamma$, β-$FeSi_2$ does not have a modulated structure but nevertheless possess a complex crystal structure due to the presence of stacking faults as reported on polycrystalline samples (Yamane et al. 2009) and also observed on thin films (Zheng et al. 1992). However, they are rarely considered in the literature, whereas they certainly have a non-negligible influence on the transport of phonons and therefore on the thermal conductivity of this material. To go further, transmission electron microscopy and electron diffraction studies were undertaken and confirmed the presence of stacking faults (Figure 8.2). This is reflected in the powder pattern by significant widening and asymmetry of given peaks which could be modeled using the FAULTS software (Casas-Cabanas et al. 2016) based on the DIFFAX algorithm (Treacy et al. 1991). In this way, it was possible to determine that the probability of the stacking faults was twice higher in the sample synthesized by MR compared to the reference samples (arc-melted), correlating the observed drop of thermal conductivity in the MR samples.

8.2.2. Synthesis of higher manganese silicide through wet ball milling

As discussed above, when produced with melting methods, $MnSi_\gamma$ always contains secondary inclusions of manganese monosilicide MnSi which is detrimental for its mechanical and transport properties. In addition, from an industrial point of view, synthesis processes with lower-energy consumption are often preferred. Therefore, processes such as mechanical alloying has been considered as a viable alternative to melting–solidification techniques. In recent years, a new process combining wet ball milling and spark plasma sintering has been proven as a simple, effective and scalable process to synthesize $MnSi_\gamma$ (Nhi Truong et al. 2015). In this study, the ball milling step was performed under both wet and dry conditions, the latter used as a reference. Indeed, mechanical alloying is usually performed under dry conditions, but in the case of HMS, the high amount of energy required for its formation leads to the agglomeration of powder. Due to such powder agglomeration, the final processed material ends up to be off-stoichiometric and the formation of detrimental MnSi secondary phase is observed (Miyazaki et al. 2018). The use of a liquid medium has the advantage of tackling the agglomeration issue during ball milling as it prevents particle welding during collisions. Additionally, it advantageously enables to decrease the final particle size. In this study (Nhi Truong et al. 2015), *n*-hexane was used for the convenient removal after ball milling as it has a low-boiling point, and also as it does not oxidize the precursor powders. Contrary to its dry ball milling counterpart, HMS powder did not contain any impurities as proven by the high quality of the X-ray synchrotron diffractograms shown in Figure 8.3. Ultimately, the measured transport properties of the wet milled sample show a 20% higher Seebeck coefficient and a 23% lower electrical resistivity than the dry milled one, over the whole measured temperature range. Moreover, the thermal conductivity was reduced by up to 30%, leading up to a high value of zT of 0.55 at 850 K for an undoped sample of $MnSi_\gamma$.

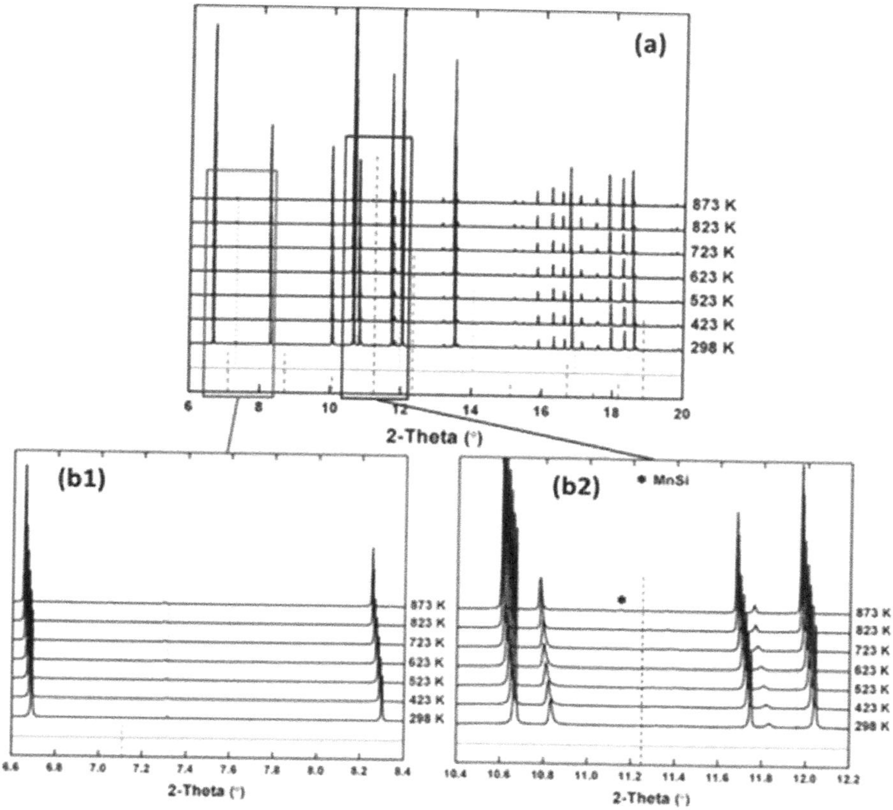

Figure 8.3. *High-temperature X-ray synchrotron diffraction pattern of a higher manganese silicide prepared by wet ball milling (ESRF, Grenoble). From Nhi Truong et al. (2015)*

8.2.3. *Issues of MnSi striations and thermal stability on thermoelectric performance of doped higher manganese silicide*

As stated above, one of the major drawbacks of HMS is its thermal instability leading to the degradation of properties. This instability is explained whether by a temperature-induced change in γ-value for samples prepared by magnesiothermy (Le Tonquesse et al. 2020b), or by the formation of MnSi secondary phase for samples synthesized through melting/solidification processes (Miyazaki et al. 2018). One way to improve the properties of HMS is by point defect and microstructure engineering through careful optimization of the carrier concentration by elemental substitution on the Mn or Si sites (Ghodke et al. 2019). Though it is indeed possible to play with the charge carrier concentration to achieve better thermoelectric properties or, for example, to tune between both *p*- and *n*-types via doping by Cr and Fe as illustrated by Figure 8.4 (Guo et al. 2021), in most cases solving the issue of formation of MnSi as a secondary phase remains a challenge.

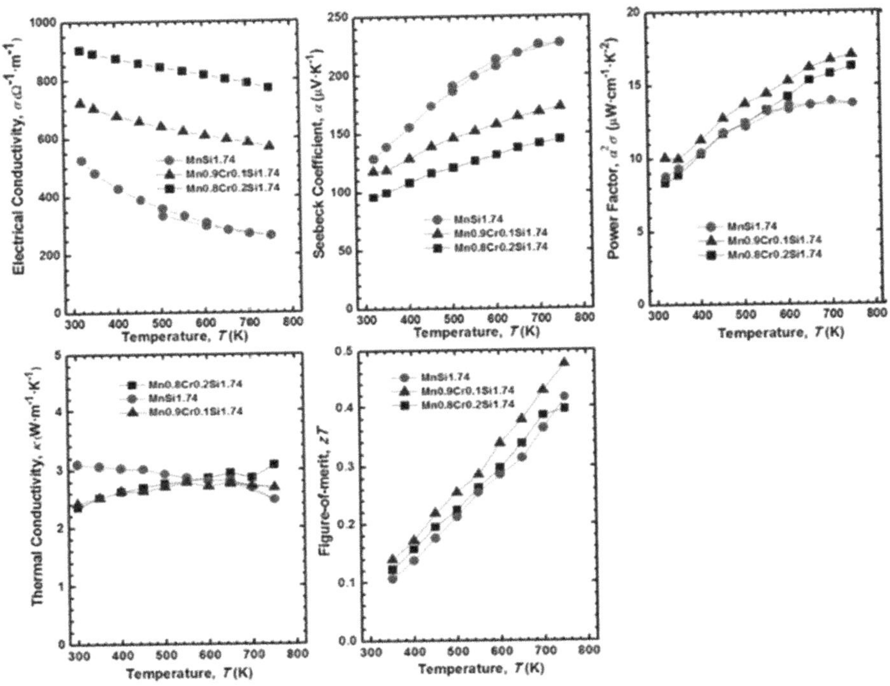

Figure 8.4. *Thermoelectric properties of undoped and Cr-doped $MnSi_{1.74}$ prepared by arc melting process followed by spark plasma sintering. From Guo et al. (2021)*

Hopefully, recent work on cobalt- (Nagai et al. 2019) or vanadium-doped HMS (Miyazaki et al. 2017) has reignited interest in solidification technics for the production of MnSiγ as it was demonstrated that, by a melt growth method a 3–4% substitution of Co or a 2% substitution of vanadium for manganese was efficient to dissipate MnSi precipitation. Indeed, as seen in Figure 8.5, X-ray diffraction patterns of the selected area of $Mn_{0.98}V_{0.02}Si_{1.74}$ melt-grown ingot show no traces of MnSi precipitation. To the exception of the top side of the ingot where accumulated silicon can be found, the remaining parts of the sample appear to be highly homogenous. These observations motivated further studies on HMS synthesis with different melting/solidification techniques.

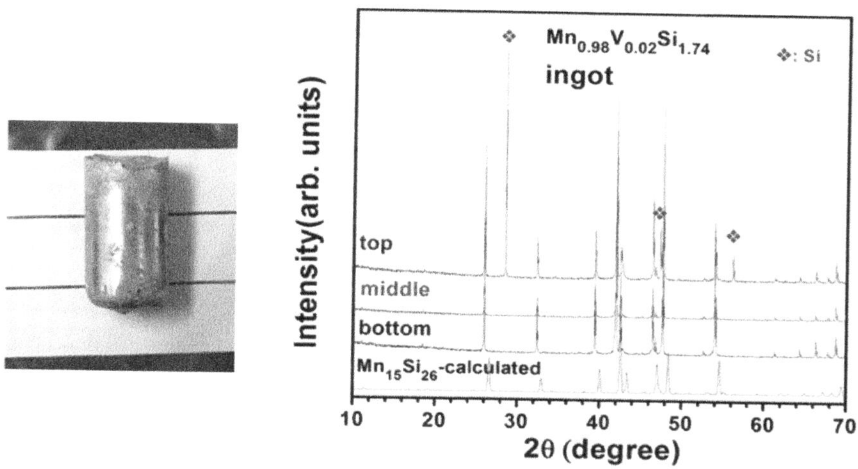

Figure 8.5. *Image and X-ray diffraction patterns of V-doped HMS sample prepared by a melt growth method (source: original results)*

Other dopants were thus considered to look for similar suppression of MnSi precipitation within the material. The highest figure of merit, zT = 1.04 at 900 K, for an HMS compound has been previously observed on Re-substituted Mn (6%) prepared by the liquid quenching technique (Yamamoto et al. 2016). Re-doped samples were thus an obvious choice to be further analyzed, and subsequent synchrotron X-ray diffraction analyses were carried out on powder samples obtained by grinding melted and rapidly quenched supersaturated Re-doped HMS (Ghodke et al. 2019). In this study, Le Bail's analyses were performed with the (3+1)-dimensional incommensurate super space group approach using JANA2006 software (Petříček et al. 2014) on both Re-free and Re-doped samples heated up to 300, 573 and 873 K. Results of these Le Bail refinements show a phase separation at high temperature for the Re-free HMS with a small amount of MnSi precipitation directly related to the change of the refined modulation vector component γ (i.e. compound stoichiometry). On the contrary, the Re-substituted HMS does not possess such a phase separation which would lead to degradation of thermoelectric performance. Consequently, and as expected, its absence leads to less significant degradation of the Re-doped HMS transport properties, in a similar way to V-doped samples.

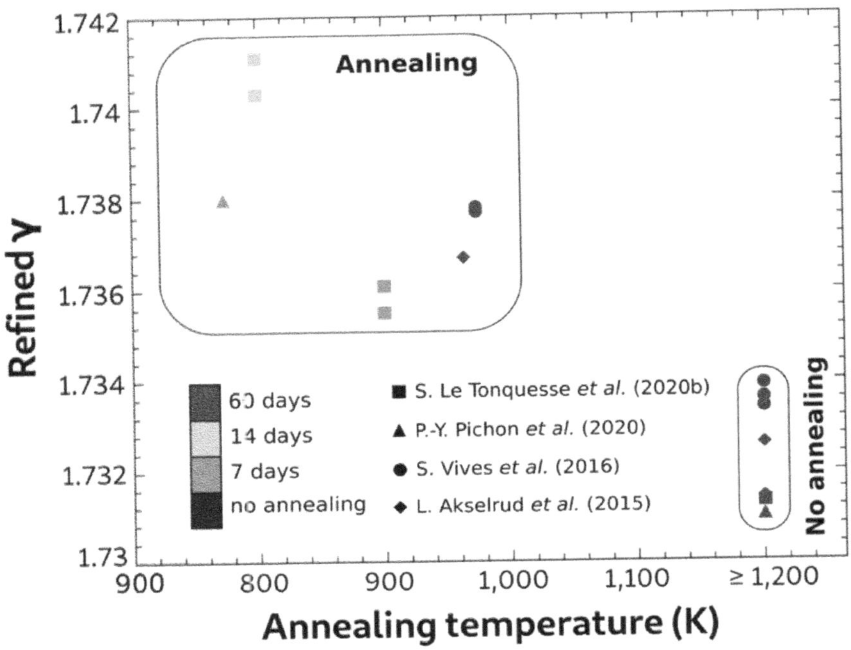

Figure 8.6. *Refined component of the modulation vector γ versus annealing conditions for HMS synthesized with different techniques. From Le Tonquesse et al. (2020b)*

Other studies performed by X-ray powder diffraction at the laboratory scale or by synchrotron radiation have also demonstrated the importance of monitoring the temperature dependence of the modulation vector component γ, whether depending on the sample composition or its thermal history. It was first exemplified by Kikuchi et al. (2014) through in situ XRD up to 1,173 K where the γ ratio was found to decrease above 773 K owing to the temperature dependence of the c_{Si} parameter and discussed as a possible cause for mechanical failure of some HMS material at high temperature. Further characterization performed by synchrotron X-ray powder diffraction with different nominal compositions and thermal history demonstrated that regardless of these parameters, all the samples will have a similar γ ~ 1.736 after an appropriate annealing time, as illustrated in Figure 8.6

(Le Tonquesse et al. 2020b). This evolution of γ depending on the thermal history of the material is therefore an important parameter to take into account before implementing this material into thermoelectric modules. In particular, deep structural characterization must be considered beforehand to ensure that the material has reached equilibrium, meaning here a stable γ. Aside of potential mechanical failure, the efficiency of the material is also at stake as it was observed in the latter study that the evolution of γ with time was followed by a power factor drop (-19%) only partially compensated by a decrease in thermal conductivity, resulting in an overall decrease of zT around -12% for zT_{max} after prolongated annealing (Le Tonquesse et al. 2020b).

8.2.4. *Upscaling processes, the examples of additive manufacturing and RGS process*

In the previous parts of this chapter, we have discussed several scalable techniques to synthesize silicide materials as well as the importance of carefully investigating their structural properties to obtain equilibrium materials with stable performances. Nevertheless, most of these techniques require a final sintering process to obtain dense materials. This is usually achieved through hot pressing or spark plasma sintering. Though these techniques are efficient and reproducible when producing materials at a lab scale, they are not always suitable to fabricate complex shape pieces or for large-scale production. Here, we will introduce recent works addressing these issues.

For example, recently, a company based in the Netherlands, RGS Development, has developed a synthesized way to produce silicide materials for thermoelectricity using their RGS (ribbon growth on substrate) technique (Pichon et al. 2020). With this solidification technique, initially developed for the production of photovoltaic materials (Lange et al. 1990; Hahn et al. 2004), the silicide materials are directly synthesized by solidification from the melt. A flat mold moving under a reservoir containing the liquid melt is used to produce sheets of silicide detachable from the mold upon cooling. This unique technique has an intermediate solidification velocity between the slow ones implying large batches (floating zones, Bridgman, etc.) and the fastest ones such as spin-casting, which allows to produce materials with unique structural and physical properties. This technique is also of great interest for industrial upscaling as it produces large quantities of thermoelectric legs with controlled dimensions in one step.

Despite its interesting features and advantages, as a non-equilibrium solidification technique, it shares the same above-described issues with other melting/solidification techniques. Indeed, most of the RGS-produced HMS were found to exhibit a temperature-dependent modulation vector component γ resulting in structural and electrical instability. As discussed before, such temperature-dependent evolution of the modulation vector component leads to MnSi precipitation in the materials and consecutively to a reduction of the charge carrier concentration and therefore of the electrical conductivity. Only for RGS-produced HMS with a 4% substitution of V for Mn does the modulation vector component γ remain temperature-independent, preventing MnSi precipitation and keeping stable electrical properties at least when the materials are kept at a temperature of up to 873 K. Finally, the 4% V-doped HMS exhibits a $zT_{max} = 0.52$ due to its thermal stability as well as because of ionized impurity scattering effects which mitigates the Seebeck coefficient reduction due to higher charge carrier concentration. This state-of-the-art value for HMS material obtained with this upscaled process is an interesting topic for future research for large-scale deployment of thermoelectric devices, though additional studies will be necessary to confirm its structural stability at higher temperature and for longer operating time.

Another potential industrial process which has recently been considered for thermoelectric silicide production is additive manufacturing using a selective laser sintering (SLS) or melting (SLM) techniques (Thimont et al. 2018). These SLS/SLM techniques are promising to produce complex-shaped legs with no requirements of machining-step. In this study, the feasibility of producing directly HMS legs through SLS/SLM is discussed. Starting from a mixture of its elemental constituents, here ball-milled manganese and silicon powders, it was demonstrated that it was possible to simultaneously obtain a $MnSi_\gamma$ phase and sinter it by additive manufacturing. This production method has many advantages, as it requires low amount of energy and does not discard materials, which would reduce the cost of leg fabrication compared to other means of production. This study is an innovative and encouraging step for future industrial applications but the method still requires further work to optimize the layer-by-layer process using SLS/SLM technique in order to fabricate any kind of leg geometries and sizes for various applications.

8.3. Towards contacts and device developments

So far, lots of efforts have been devoted to the synthesis and process development of transition metal silicide materials but the realization of functional devices integrating them still remains a challenge. Some key limitations to produce devices made of silicide materials are (i) compatible thermal expansion coefficients (TEC) between *n*- and *p*-type materials, and (ii) the realization of good-quality contacts and diffusion barriers. In the past, performances of "all-silicide-modules" have already been evaluated (Skomedal et al. 2016). In this study, a module made of HMS as *p*-type and Sn-doped Mg_2Si as *n*-type were designed and tested. In order to overcome the issue of large TEC differences between both materials, the module was designed in a way to integrate spring-loaded contacts on the cold side to mitigate the TEC mismatch. Despite achieving a respectable efficiency of up to 5%, the module tested above 673 K shows significant and irreversible damage due to oxidation and diffusion issues between the materials and the metallization. Anyway, this study demonstrated the possibility to achieve a decent power output and efficiency with a module made up of low-cost and abundant materials but that still needs improvement of oxidation resistance and long-term stability of the devices.

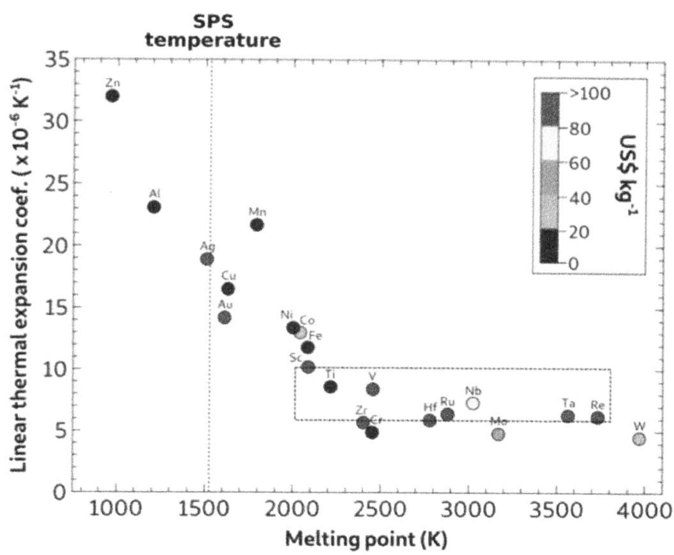

Figure 8.7. *Linear thermal expansion coefficient versus melting point of selected potential contact metals for direct sintering with $CrSi_2$*

In that regard, recent work was focused on the realization of metal contacts on $CrSi_2$ in order to develop stable and efficient thermoelectric legs (Le Tonquesse et al. 2021). Relatively earth-abundant and nontoxic Nb and Ti were selected as potential contact materials for their compatible TEC with $CrSi_2$ (Figure 8.7). Preliminary high-temperature XRD measurements were carried out to carefully assess the $CrSi_2$ TEC. Sandwiched Ti/$CrSi_2$/Ti and Nb/$CrSi_2$/Nb legs were then directly synthesized through spark plasma sintering. The Nb contacts demonstrated poor adhesion and metallic properties, while the Ti contacts were shown to have limited diffusion at the interface and good mechanical resistance. Nb contact failure was attributed to sample internal stress due to larger TEC mismatch and also to significant silicon diffusion into niobium. The Ti/$CrSi_2$/Ti legs exhibited low contact resistance and a moderate efficiency, but most importantly, this study demonstrated that it was possible to produce good-quality metallic contacts on low-cost transition metal silicide materials with stable mechanical and chemical properties. This work paves the way to realize more efficient transition metals silicide legs towards more complex architectures and more elaborate diffusion barriers.

8.4. References

Casas-Cabanas, M., Reynaud, M., Rikarte, J., Horbach, P., Rodriguez-Carvajal, J., (2016). FAULTS: A program for refinement of structures with extended defects. *J. Appl. Crystallogr.*, 49, 2259–2269.

Ghodke, S., Sobota, R., Berthebaud, D., Pichon, P.-Y., Navone, C., Takeuchi, T. (2019). Effect of re substitution on the phase stability of complex $MnSi_\gamma$. *J. Electron. Mater.*, 48, 5827–5834.

Guo, Q., Zhang, W., Liu, Z., Fu, X., Le Tonquesse, S., Sato, N., Son, H-W., Shimamura, K., Berthebaud, D., Mori, T. (2021). Thermoelectric performance of Cr doped and Cr–Fe double-doped higher manganese silicides with adjusted carrier concentration and significant electron–phonon interaction. *ACS Appl. Mater. Interfaces*, 13, 8574–8583.

Hahn, G. and Schönecker, A. (2004). New crystalline silicon ribbon materials for photovoltaics. *J. Phys. Condens. Matter*, 16, 1615–1648.

Kikuchi, Y., Nakajo, T., Hayashi, K., Miyazaki, Y. (2014). High temperature X-ray diffraction study on incommensurate composite crystal $MnSi_\gamma$ – (3+1)-dimensional superspace approach. *J. Alloys Compd.*, 616, 263–267.

Lange, H. and Schwirtlich, I.A. (1990). Ribbon growth on substrate (RGS) – A new approach to high speed growth of silicon ribbons for photovoltaics. *J. Cryst. Growth*, 104, 108–112.

Le Tonquesse, S., Verastegui, Z., Huynh, H., Dorcet, V., Guo, Q., Demange V., Prestipino, C., Berthebaud, D., Mori, T., Pasturel, M. (2019). Magnesioreduction synthesis of Co-Doped beta-$FeSi_2$: Mechanism, microstructure, and improved thermoelectric properties. *ACS Appl. Energy Mater.*, 2, 8525–8534.

Le Tonquesse, S., Dorcet, V., Joanny, L., Demange, V., Prestipino, C., Guo, Q., Berthebaud, D., Mori, T., Pasturel, M. (2020a). Mesostructure-thermoelectric properties relationships in $V_xMn_{1-x}Si_{1.74}$ (x = 0, 0.04) higher manganese silicides prepared by magnesiothermy. *J. Alloys Compd.*, 816, 152577.

Le Tonquesse, S., Joanny, L., Guo, Q., Elkaim, E., Demange, V., Berthebaud, D., Mori, T., Pasturel, M., Prestipino, C. (2020b). Influence of stoichiometry and aging at operating temperature on thermoelectric higher manganese silicides. *Chem. Mater.*, 32, 10601–10609.

Le Tonquesse, S., Matsushita, Y., Jood, P., Ohta, M., Mori, T., Berthebaud, D. (2021). Fabrication and evaluation of low-cost $CrSi_2$ thermoelectric legs. *Crystals*, 11, 1140.

Miyazaki, Y., Igarashi, D., Hayashi, K., Kajitani, T., Yubuta, K. (2008). Modulated crystal structure of chimney-ladder higher manganese silicides $MnSi_\gamma$ (γ~1.74). *Phys. Rev. B*, 78, 214104.

Miyazaki, Y., Hamada, H., Hayashi, K., Yubuta, K. (2017). Crystal structure and thermoelectric properties of lightly vanadium-substituted higher manganese silicides $(Mn_{1-x}V_x)Si_\gamma$. *J. Electron. Mater.*, 46, 2705–2709.

Miyazaki, Y., Hamada, H., Nagai, H., Hayashi, K. (2018). Crystal structure and thermoelectric properties of lightly substituted higher manganese silicides. *Materials*, 11, 926.

Nagai, H., Hamada, H., Hayashi, K., Miyazaki, Y. (2019). Effects of cobalt substitution on crystal structure and thermoelectric properties of melt-grown higher manganese silicides. *J. Electron. Mater.*, 48, 1902–1908.

Nhi Truong, D.Y. (2015). Thermoelectric properties of higher manganese silicides. PhD Thesis, University of Waterloo, Waterloo.

Nhi Truong, D.Y., Kleinke, H., Gascoin, F. (2015). Preparation of pure higher manganese silicides through wet ball milling and reactive sintering with enhanced thermoelectric properties. *Intermetallics*, 66, 127–132.

Petříček, V., Dušek, M., Palatinus, L. (2014). Crystallographic computing system JANA2006: General features. *Z. Krist. – Cryst. Mater.*, 229, 345.

Pichon, P.-Y., Berneron, P., Levinsky, J., Burema, A., Blake, G., Berthebaud, D., Gascoin, S., Gascoin, F., Hébert, S., Amtsfeld, J. et al. (2020). Stability and thermoelectric performance of doped higher manganese silicide materials solidified by RGS (ribbon growth on substrate) synthesis. *J. Alloys Compd.*, 832, 154602.

Rodriguez-Carvajal, J. (1993). Recent advances in magnetic-structure determination by neutron powder diffraction. *Physica B*, 192, 55–69.

Skomedal, G., Holmgren, L., Middleton, H., Eremin, I.S., Isachenko, G.N., Jaegle, M., Tarantik, K., Vlachos, N., Manoli, M., Kyratsi, T. et al. (2016). Design, assembly and characterization of silicide-based thermoelectric modules. *Energy Convers. Manag.*, 110, 13–21.

Thimont, Y., Presmanes, L., Baylac, V., Tailhades, P., Berthebaud, D., Gascoin, F. (2018). Thermoelectric higher manganese silicide: Synthetized, sintered and shaped simultaneously by selective laser sintering/melting additive manufacturing technique. *Mater. Lett.*, 214, 236–239.

Treacy, M.M.J., Newsam, J.M., Deem, N.W. (1991). A general recursion method for calculating diffracted intensities from crystals containing planar faults. *Proc. R. Soc. London Ser. A*, 433, 499–520.

Yamamoto, A., Ghodke, S., Miyazaki, H., Inukai, M., Nishino, Y., Matsunami, M., Takeuchi, T. (2016). Thermoelectric properties of supersaturated Re solid solution of higher manganese silicides. *Jpn. J. Appl. Phys.*, 55, 020301.

Yamane, H. and Yamada, T. (2009). Effects of stacking fault on the diffraction intensities of β-$FeSi_2$. *J. Alloys Compd.*, 476(1–2), 282–287.

Zheng, Y., Taccoen, A., Petroff, J.F. (1992). Planar defects in β-iron disilicide (β-$FeSi_2$) analyzed by transmission electron-microscopy and modeling, *J. Appl. Crystallogr.*, 25, 122–128.

9

Application of the Thermoelectrics; Past, Present and Future

Hirokuni HACHIUMA
Komatsu Ltd., Japan

9.1. Introduction

Thermoelectric conversion is a technology that was discovered in the 19th century. It consists of two effects; the Seebeck effect, which makes electricity from temperature differences, and the Peltier effect, which makes temperature differences from electricity. In the mid-19th century, various thermoelectric generators were tried. These can be found in documents of that time and museums, including a patent from 1870. However, due to its low efficiency, it was never widely used. The Peltier effect was not applied until the 1950s when a paper was published on BiTe to produce temperatures below 0°C. BiTe is still the main and almost the only practical material used in thermoelectric applications. The primary purpose of thermoelectric conversion today is in cooling and temperature control. Its products are commonly named Peltier devices, and it has become an indispensable technology in various fields, from consumer to research and industrial applications. However, with the growing awareness of energy-saving and

For a color version of all the figures in this chapter, see www.iste.co.uk/akinaga/thermoelectric2.zip.

Thermoelectric Micro/Nano Generators 2,
coordinated by Hiroyuki AKINAGA, Atsuko KOSUGA,
Takao MORI and Gustavo ARDILA.

carbon neutrality in recent years, there have been many attempts to recover electricity from the waste heat of factories or engines using thermoelectrics. It is also attracting attention as an energy-harvesting power source for IoT devices, i.e. as a technology to generate power from minute heat flows in the environment.

Figure 9.1 shows the number of patent applications related to thermoelectricity in Japan. There was a peak in the 1950s and 1960s when Peltier devices were in their beginning stage, and the following rise was during the IT bubble period around 2000. The increase after 2010 reflects projects targeting power generation applications. Thermoelectric application products are niche, but their market and applications are steadily increasing.

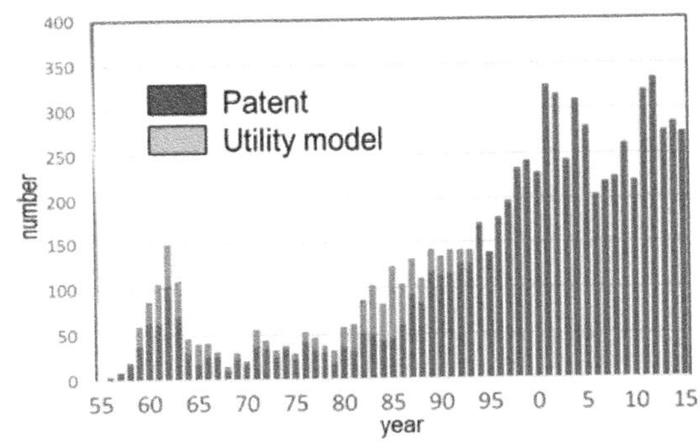

Figure 9.1. *Number of patent applications in Japan*

9.2. Thermoelectric module

A thermoelectric module is constructed by connecting p-type and n-type thermoelectric elements as in Figure 9.2. The elements are connected electrically in series and thermally in parallel. One pair of p and n elements is a minimum set, and the performance relies on the number of pairs. The conversion efficiency of an element is almost the same if the cross-section and length aspect ratio are the same. As is the case with other electronic components, the smaller the size, the greater its relative advantage over other technologies because of the size effect.

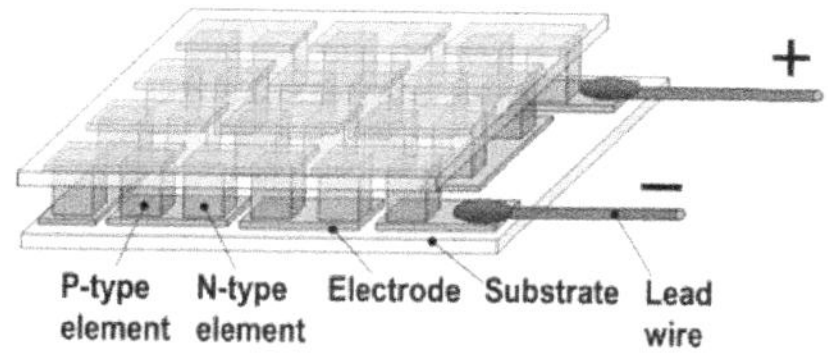

Figure 9.2. *Structure of a thermoelectric module*

A heat exchanger is necessary to transfer heat to and from the thermoelectric module. It is designed to match the form of target heat and is attached to both sides of the module surface. For the best performance, the module and heat exchanger must be designed as a single unit.

This chapter presents several examples of products that use thermoelectric conversion.

9.3. TEC application for refrigerator and cooler

It is no surprise that refrigerators and air conditioners were the first to use TEC, thermoelectric cooling modules to control temperature. RCA announced a prototype of a small refrigerator using thermoelectric panels in 1954. In 1958, KELK developed a prototype of Japan's first thermoelectric refrigerator. In addition, an article in 1959 reported Westinghouse was converting its 250,000-unit-per-year compressor refrigerator plant into a thermoelectric refrigerator one.

It is no wonder that people thought this was a miracle technology to replace the conventional refrigerator since it has a straightforward structure, just inserting thermoelectric cooling panels.

The KELK's refrigerator, shown in Figure 9.3, was water-cooled, and the thermoelectric cooling panel used in the refrigerator consisted of 42 pairs of elements. When the refrigerator was empty, it was able to lower the internal temperature to -6°C after two hours at an outdoor air temperature of 20°C and to -2°C at 30°C. When the refrigerator was operated with 3 liters of water, the internal temperature was lowered to 10°C after two hours at an outdoor air temperature of 20°C, and to 4°C after five hours. We confirmed that it could function as a refrigerator.

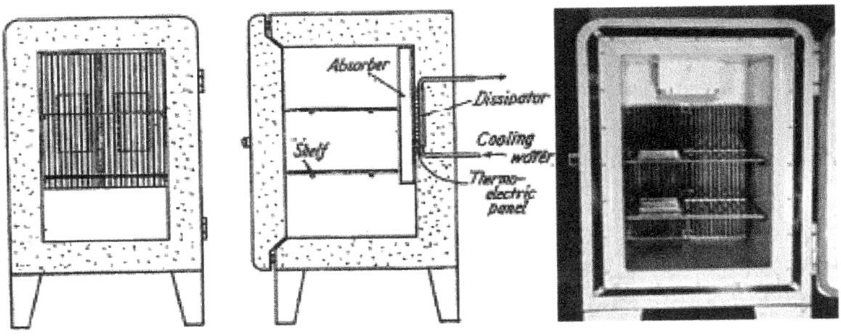

Figure 9.3. *Prototype of a thermoelectric refrigerator in 1958*

However, it had low efficiency. The power consumption was seven to eight times higher than a compressor refrigerator. The performance of the element material used in the prototype was lower. Still, even if we had used the current best material, the power consumption would have been about four to five times higher. Also, although the structure of the thermoelectric cooling panel was supposed to be simple and low cost, the manufacturing cost was expensive due to the use of rare metals as materials.

From these failures, people have learned the following features of thermoelectricity and developed valuable products:

1) Design specifications can be easily adjusted by simply determining the number of pairs to achieve the required performance, and the efficiency does not decrease even if it is made smaller.

2) Cooling capacity can be easily adjusted by the electric current value. It works for both cooling and heating depending on the current direction.

3) There are no moving parts, so no vibration or noise.

If you can design a product that takes advantage of the above, it will become a product that cannot be realized with other technologies, despite its niche.

A good example is a wine cellar. Thermoelectric devices are ideal for wines that lose their taste due to vibration. In addition, the ideal temperature for wine storage is 14°C, so thermoelectric devices can fully show their performance.

Another example is a noiseless refrigerator for hotels and hospitals. The operating noise of compressor refrigerators is a significant problem in hotels and hospitals, where the rooms are small and intended only for sleeping. The capacity is small for personal use, and just a small performance is enough as the only thing needed is to keep the temperature inside.

Figure 9.4 shows a prototype of an in-vehicle thermoelectric air cooler developed by KELK in 1958. However, this prototype did not have the performance to cool the entire car or make the driver comfortable. Today, car seat air conditioners that can dry and cool the sweaty areas around the back and legs are used in high-end cars. Recently, many consumer goods with a thermoelectric module for personal comfort have been on the market that provide cooling for the back and neck or air-conditioned clothing. These successful applications do not require a big cooling capacity. The ability to cool a small area in a short period suits the characteristics of thermoelectricity, rather than cooling a large volume of air.

Figure 9.4. *Prototype of a thermoelectric car cooler in 1958*

Overcoming past failures, a large part of the thermoelectric market is for consumer products such as refrigerators and coolers, and it has a high growth rate. Figure 9.5 shows the thermoelectric market as of 2004 and 2014.

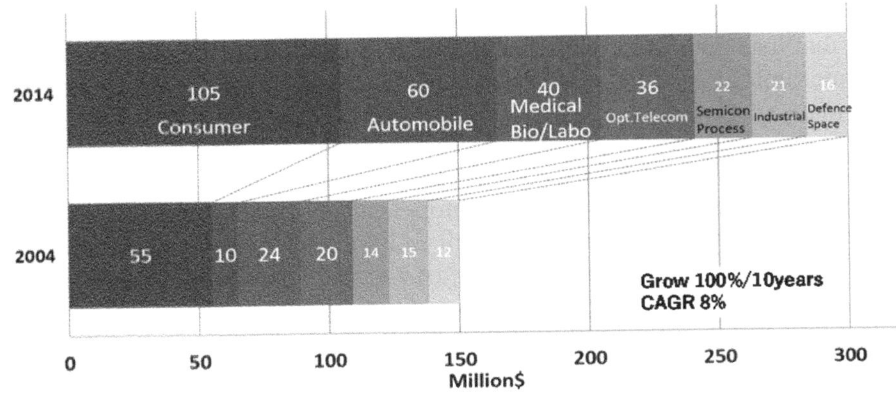

Figure 9.5. *Market size of TEC*

9.4. TEC for electronic components

Electronic components have become smaller as they have evolved. The application for electronic devices that cool a small area is the best application that only a TEC, thermoelectric cooling module can provide.

Localized hot spots appear when operating microelectronic circuits. The smaller the device, the greater the heat density, and the more cooling is required to prevent degradation of the device. In particular, the characteristics of optical devices are affected by temperature fluctuations. Therefore, precise temperature control is required for applications that demand advanced performance.

The use of cooling computer CPUs for improving performance is known to mechanics. Data centers use air conditioning power equivalent to that of the server machines to cool the entire room and remove the heat generated mainly by the high-performance CPU, so that the hot spots do not exceed the deterioration temperature. If a TEC module can eliminate the hot spots, it may be possible to lower the air conditioning energy.

9.4.1. *TEC for optical communication*

In terms of applications in electronic devices, the largest market is for cooling lasers used in optical communications. Figure 9.6 shows the structure of the TEC-mounted laser package. Two critical technologies are needed to implement the worldwide installation of optical communication networks and the spread of the Internet. One is DWDM (Wavelength Division Multiplexing), a technology that expands the bandwidth by transmitting signals of many wavelengths simultaneously through a single optical fiber. Another is EDFA (Erbium-Doped Fiber Amplifier), a technology that transmits signals for several hundred kilometers or more using an optical laser signal amplifier.

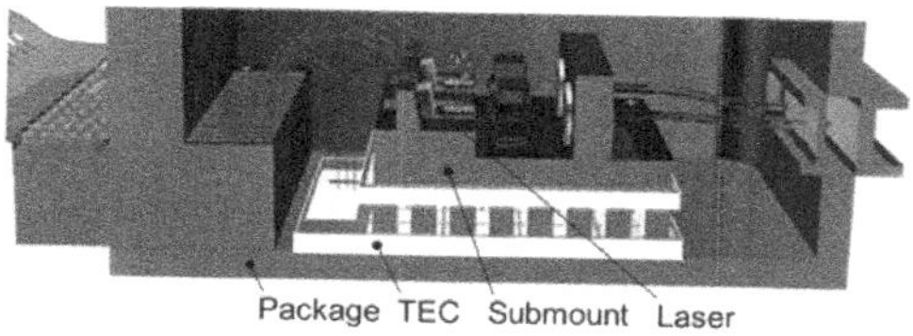

Figure 9.6. *Structure of TEC-mounted laser component for optical communication*

The wavelength of light emitted from a semiconductor laser varies with the temperature. The length of the laser oscillating semiconductor crystal expands as the temperature rises, resulting in an increase in the resonant distance and an increase in wavelength. Figure 9.7 shows the properties. The wavelength varies by 0.3 nm with a temperature change of 1 degree. The optical signal degrades in fiber, through attenuation and the dispersion of pulses. The effects depend on the wavelength, and there are wavelength bands where these degradations are the weakest, which are the most favorable for transmission. The optical wavelength band range suitable for the DWDM source laser is only about 30 nm, approximately 1,550 nm. If there are 100 wavelengths in 30 nm, the wavelength gap is 0.3 nm, so if the temperature changes by 1 degree, the signals will overlap. Therefore, it is necessary to control the temperature of the source laser for DWDM in the order of 0.1 degrees using a TEC module.

The pump laser used in EDFA does not require precise temperature control. But its laser power is more than 10 times greater than that of the source laser. The temperature of the laser cannot be maintained low enough unless it is actively cooled so that the device will not deteriorate. Figure 9.8 shows an example of the correlation between temperature and the lifetime of a semiconductor laser. If the temperature rises by 20 degrees, the lifetime drops to 1/10. For this reason, the temperature of the laser is usually controlled between 25 and 35°C.

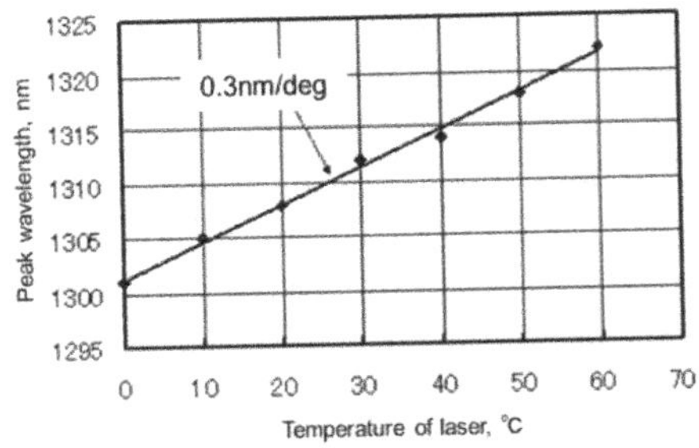

Figure 9.7. *Effect of temperature on semiconductor laser wavelength*

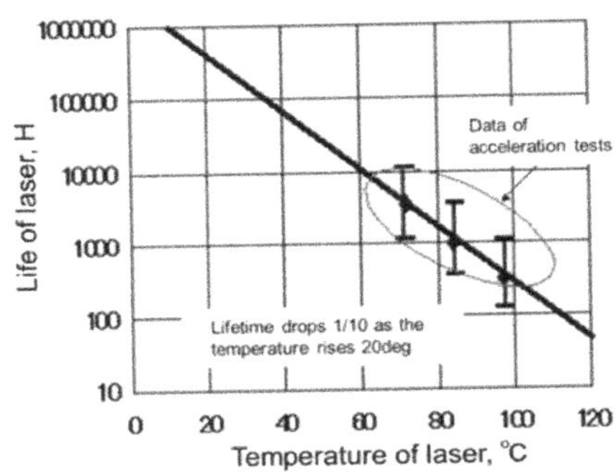

Figure 9.8. *Effect of temperature on semiconductor laser life*

9.4.2. *Multi-stage TEC for optical sensors*

A single-stage thermoelectric cooling module has a maximum temperature difference of 65–75°C when the hot side temperature is 27°C. A multi-stage module can produce a larger temperature difference by stacking these modules and lowering the temperature in each stage. For example, the maximum temperature difference of a four-stage module is 127 degrees and that of a five-stage module is 140 degrees. Cold-side temperature is -100°C for four stage and -113°C for five stage. These are conditions with no thermal load. They usually have a thermal load of several tens to several hundred mW, and the cold side temperature is 10 degrees higher.

Figure 9.9 shows the status of the five-stage module during operation. The heat absorption of each stage is equal to the heat absorption of its upper stage plus the power consumption. Therefore, the lower the stage, the more heat absorption, and the lower the stage, the more the number of element pairs. Also, during operation, the COP of the multi-stage module is extremely low. The multi-stage module sacrifices this because its function is to produce a high ΔT.

stage	pairs
5	2
4	5
3	11
2	26
1	64

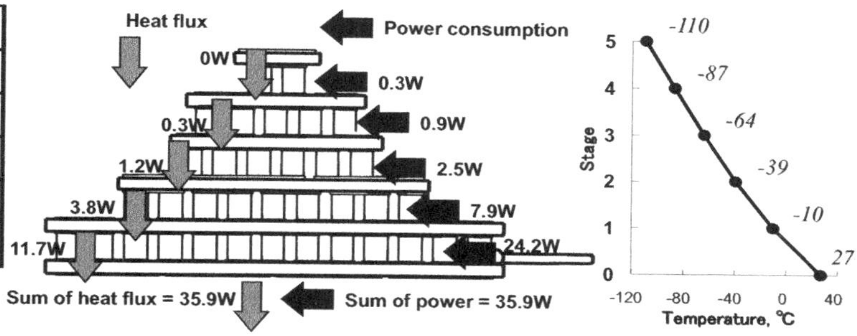

Figure 9.9. *Status of the five-stage module during operation under no heat load*

Infrared sensors and high-performance optical sensors are the applications using the multi-stage module. The following is a description of these applications.

An infrared sensor detects light in the infrared region. It is used for security monitoring equipment and thermal imaging cameras. Quantum infrared sensors have high sensitivity and excellent response time, but high sensor temperature lowers the S/N ratio and causes problems in signal detection. Therefore, it needs to be cooled with some method. Figure 9.10 shows the relationship between the temperature and spectral response characteristics of a PbSe infrared sensor. By cooling the sensor from 25°C to -20°C, the sensitivity becomes four times higher.

High-sensitivity, long-exposure image sensors for low-light applications such as fluorescence microscopy, astronomical observation and spectroscopic measurement use a thermoelectric cooling module. The dark current that reduces the S/N ratio in an image sensor is caused by thermal noise, and cooling can reduce it. Figure 9.11 shows an example of the dark current characteristics of an image sensor. Cooling reduces the dark current exponentially. In the low-temperature range, the dark current is reduced by half when the temperature drops by 5°C.

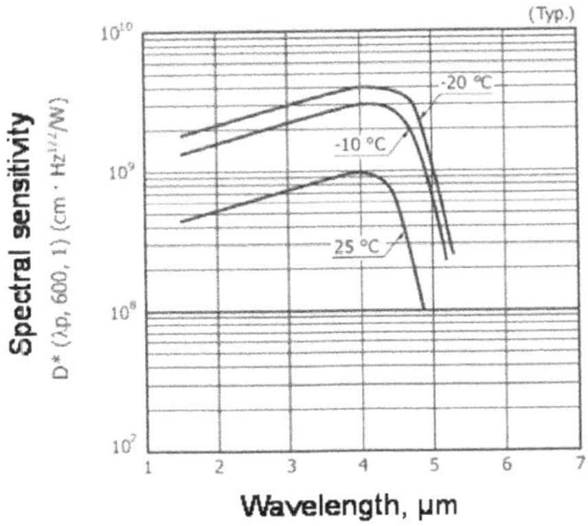

Figure 9.10. *Temperature and sensitivity of an infrared sensor*

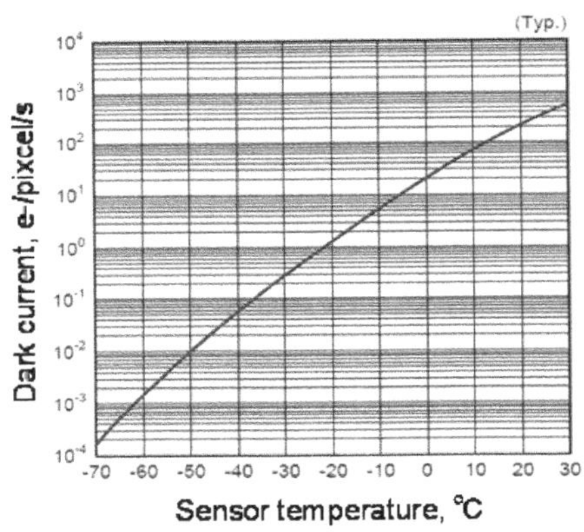

Figure 9.11. *Temperature and dark current of an image sensor*

These photosensors are passive elements, and power consumption is extremely low. The purpose of localized ultra-low temperatures for only small sensors is an ideal application to take advantage of the characteristics of multi-stage TEC. Using a thermoelectric module (1) eliminates vibrations compared to a mechanical refrigerator and enables high-definition images without interfering with a precise observation by a microscope or telescope, and (2) enables local cooling of the image sensor without condensation on the peripheral equipment. These features of the thermoelectric cooling module are helpful in long exposure applications.

9.5. TEC for semiconductor manufacturing

TEC integrated temperature control systems have been widely applied in various semiconductor manufacturing processes. In the semiconductor manufacturing process, a precise temperature control of about ±0.1°C near room temperature is required, and this is a field where the features of thermoelectric cooling modules are advantageous.

Following Moore's law, semiconductor products have achieved higher performance and lower cost year after year by increasing the integration of more circuits of the same size. To achieve narrower circuits on silicon wafers, eliminating the fluctuations in the process conditions is essential. Therefore, it has become necessary to precisely control the temperature of the resist, etching and other chemicals, cleaning chemicals, and even the silicon wafer itself. Figures 9.12 and 9.13 show the TEC integrated heat exchanger for liquid temperature control and silicon wafer temperature control, respectively.

Figure 9.12. *A heat exchanger for liquid*

Figure 9.13. *A wafer cooling plate*

In the semiconductor manufacturing lab, conventional temperature control systems for liquid are oversized. They are installed in utility rooms rather than cleanrooms due to the problems of dust emission and vibration from moving parts. As shown in Figure 9.14, the thermoelectric temperature control system is compact and can be installed near the semiconductor manufacturing equipment, and has many advantages:

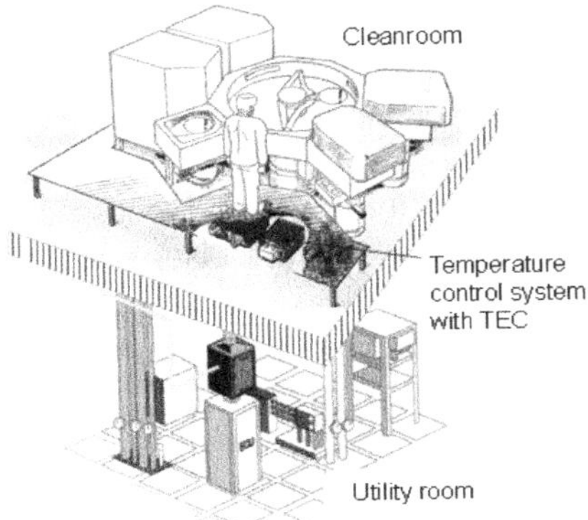

Figure 9.14. *Point-of-use concept by small system with TEC*

1) Temperature control accuracy

The temperature accuracy of the thermoelectric temperature control system is less than ±0.1°C. The manufacturing conditions of around or above room temperature are best suited for using a thermoelectric module as it controls both cooling and heating by simply switching the current direction. In addition, by placing the thermoelectric temperature control system close to the manufacturing equipment, feedback control can be made without time delays in response to temperature fluctuations, resulting in improved control accuracy.

2) Energy-saving

Due to its location close to the equipment, there is less heat loss and pressure drop in the piping, resulting in energy savings. Also, conventional systems use a compressor for cooling and a combined heater, resulting in wasted energy, while the thermoelectric temperature control systems save energy.

3) No vibration and dust

Since there are no moving parts, no dust or vibration can harm the process, even if the system is installed near the equipment in a cleanroom.

4) Compact and space-saving

The product is compact and can be installed in a dead space inside the equipment, reducing the installation area to almost zero.

5) CFC-free

The thermoelectrics make it environmentally friendly and CFC-free.

9.6. TEG application

So far, the application of TEG, thermoelectric power generation, is limited to extraordinary usage. One example is an RTG or radioisotope thermoelectric generator, which uses the decay heat of radioactive material as a heat source. Another is the anticorrosive protection of pipelines, which use gas or oil for fuel. Both applications are used in an extremely remote area, far away from the power infrastructure. Therefore, reliability is the most important, and the cost is not a big issue.

In order to use TEG not only for special applications, but also for more general use, their cost performance needs to be improved; during the oil crisis of the 1970s and 1980s, industry, government and academia actively conducted research and development of TEG. However, as the energy cost decreased, the activity slowed down, and the practical application was not realized. In recent years, the potential of TEG has been attracting renewed attention due to the improvement of peripheral technologies such as power management technology and low power consumption technology, as well as the growing interest in global warming and carbon neutrality.

Practical application projects of TEG have been successively carried out in Japan, the USA and Europe since around 2000. The recovery of industrial waste heat was the central issue of the project in Japan, and automobile waste heat recovery was the central issue in the USA and Europe. Though it is not practical yet, the TEG development project has been developed more widely and deeply. Several examples are shown below.

TEG can be primarily divided into three categories, as shown in Figure 9.15, according to the heat source scale and electric power purpose. (1) Energy harvesting applications for sensors, actuators and wireless communications that operate with micropower by capturing minute heat flows that exist in the environment and converting them into mW class electric power. (2) Independent power supply applications that convert the heat of few kW such as open fires and stoves into W class electric power for local use or charging portable equipment. (3) Waste heat recovery application over kW class from industrial waste heat for energy saving. There are existing power sources in each category, and thermoelectric power generation is considered to replace them with added value.

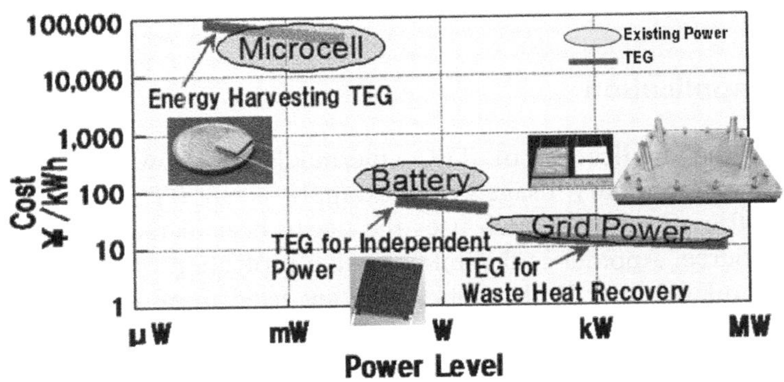

Figure 9.15. *Three categories of TEG*

9.6.1. *TEG for energy harvesting (EH)*

With the development of energy-saving technologies in recent years, EH is regarded as a technology to monitor machines without a power supply. The application best matches the thermoelectric features, "small".

The thermoelectric module for EH, which is about 5 mm the size, can produce 100 μW only by adding a temperature difference of 2–3 degrees. There are many places where this level of temperature difference appears temporarily. The machine in operation is where heat flows continuously, and enough temperature differences appear. Therefore, thermoelectricity is the most promising technology around machines among several EH technologies.

In general, after the period of initial failure, mechanical components enter a period of accidental failure and eventually reach the end of their life. Time-based maintenance (TBM), which is currently the mainstream method, is sufficient to a certain extent for preventive maintenance. Still, it is impossible to set an appropriate maintenance period to prevent accidental failures that occur randomly for each piece of equipment. Therefore, condition-based maintenance (CBM) is becoming increasingly important in recent factory management as competition intensifies. In CBM, sensors for temperature, pressure, vibration, etc., need to be installed at the monitoring location and data needs to be acquired continuously. However, the operation of CBM has been limited to large-scale plants such as steel mills due to the high cost of sensors, their wiring, data collection/decision systems, operation and maintenance.

The "thermoelectric EH wireless sensor device" shown in Figure 9.16 is expected to solve this problem for any machine. Moreover, since it is wireless and easy to attach and remove, it is possible and suitable to install on the existing machines.

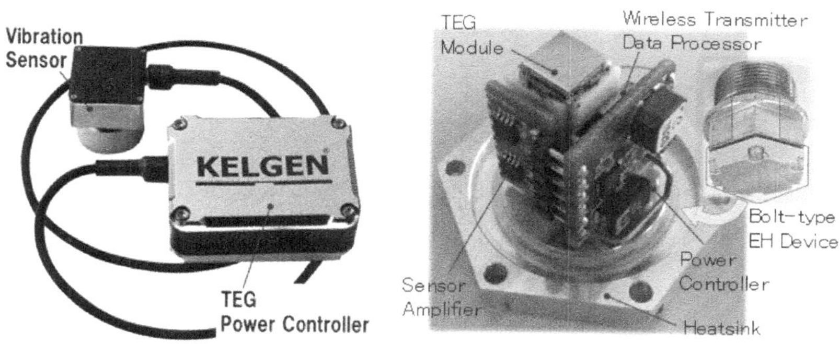

a) Vibration sensor b) Bolt type sensor

Figure 9.16. *Examples of TEG EH wireless sensor device*

The transmission interval of the EH wireless sensor device depends on the power generation quantity of the micro-TEG. The interval is several seconds to several minutes if the power generation is of a few mW to tens of μW, which is available in the TEG with a temperature difference of 30–40 degrees to 7–8 degrees between the machine surface and ambient air. Since the voltage of the thermoelectric module is low at such a small temperature difference, a booster circuit is necessary, and its characteristic matching with TEG is an essential point of the electric design.

Since a machine generates stable heat during operation, the heat flux through the TEG is also steady, and the power generation quantity is constant, even if the ambient changes. Temperature difference on the device is kept unchanging, even when the temperature fluctuates almost 10 degrees, or suddenly drops due to the rain. The data are transmitted at almost consistent intervals.

Figure 9.17 shows PoC examples of vibration monitoring for a motor outside the factory and a construction machine. TEG EH devices' power-less and wireless features are valuable in places where wiring is complicated because of moving machines, and where it is difficult to replace batteries such as in high places or remote areas.

Figure 9.17. *PoCs of TEG EH sensors*

9.6.2. *TEG for stand-alone power source*

The TEG module works almost permanently as a stand-alone power source without any special wiring if there is a heat source. A small amount of flame level thermal energy can substitute several to tens of W batteries. Not only does it has a value of no wiring and no battery, but it is also cheaper than batteries in terms of energy cost. If the flame can produce electric power while performing functions such as heating and cooking by designing well, the power generation will be a pure added value. The applied products like cartridge gas fan heaters, stoves, outdoor goods and so on are already commercially available.

The "stand-alone power unit with TEG" shown in Figure 9.18 is a product that assembles a TEG module with a heat receiving plate, a DC–DC converter, a fin and a fan. The heat receiving side is heated to about 220°C. The fan cools the heat-dissipating side to maintain the temperature difference, and a DC output of several W is possible.

Figure 9.18. *Stand-alone power unit*

9.6.3. *TEG for waste heat recovery*

The industrial field uses more than 1/3 of the primary energy, and the waste heat is large in quantity and too good in quality as the temperature is high. Energy saving in the industrial field has been pursued from economic motivation for efficiency improvement. Still, the recovery of waste heat is not done enough. Industrial furnaces, which generate a large amount of waste heat, usually operate continuously for 24 hours to prevent heat loss due to starting and stopping, with high efficiency. For the power generation equipment of waste heat recovery, it is a field with the possibility of high operation rate and large power generation.

Figure 9.19 shows the verification test of a "TEG waste heat recovery unit" on the exhaust heat from the afterburner of a carburizing furnace. The power output is connected to the electric utility's distribution system. The system was certified as a power generation facility, on authorization from METI, for the first time in Japan as a TEG. The heat amount of a general carburizing furnace's afterburner is about several tens of kW. The TEG produces several hundred W when the total power conversion rate is a few %. Of course, it is a small output, but as the waste heat is scattered, TEG seems effective for such an application.

Figure 9.19. *Waste heat recovery from the afterburner of a furnace*

In the heat treatment process, the flame of the carburizing furnace afterburner is not constant. There are fluctuations in its thermal quantity. It would cause fatigue degradation from thermal stress. In addition, there is temperature distribution on the heat receiving plate, and local degradation is a risk. Though the life equivalent to 10 years has been confirmed in the TEG module specification temperature, the durability for the whole system is a problem in actual use.

Figure 9.20 shows a verification test using radiant heat from a continuous casting facility in steel works. The iron and steel industry is a high energy-consuming industry that occupies about 10% of the energy consumption of the whole of Japan. About 1/3 of its energy is unused and wasted as heat. Although the steel industry in Japan overwhelms other countries in energy efficiency, a huge amount of waste heat is still not used. Since radiant heat is not easy to collect as it diffuses from the site, efficient waste heat recovery is difficult for the conventional turbine generator.

On the other hand, the TEG waste heat recovery unit is advantageous in layout because it can directly generate electricity by receiving radiation heat on the heat receiving plate. In this verification test, 56 units are set in an area of 2 m width and 4 m length, and a power generation of 10 kW is realized. Even in the steel industry, where energy saving and waste heat recovery are completely done, thermoelectric technology could be applied for the small amount of wasted heat remaining.

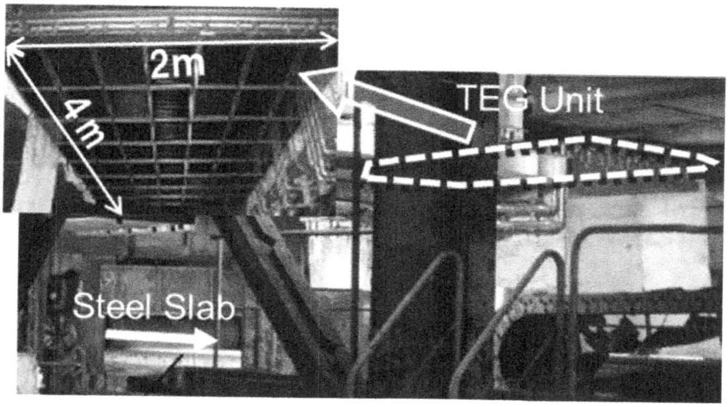

Figure 9.20. *Waste heat recovery from continuous casting in steel works*

9.7. Conclusion

Thermoelectric technology has a long history. For TEC applications, it has become a niche but indispensable technology that has been industrialized and growing for decades. People now understand how to design products that capture the features of thermoelectricity. The applications of TEG are still in the PoC stage. But if you learn and make good use of the characteristics of thermoelectricity, it will be an indispensable technology and a steadily growing industry, as is TEC.

9.8. References

Goldsmid, H.J. and Douglas, R.H. (1954). The use of semiconductors in thermoelectric refrigeration. *Brit. J. Appl. Phys.*, 5, 386–390.

Hachiuma. H. and Fukuda, K. (2007). Activities and future vision of komatsu thermo modules. *Proc. 5th European Conference on Thermoelectrics*, ECT2007.

Hachiuma, H., Murase, T., Goto, D., Fujimoto, S., Makino, K. (2019). Thermoelectric generation technology and the applied products. *Komatsu Technical Report*, 64(171), 47–53.

Hartman, R.F. (1995). Modular RTG technology. In *CRC Handbook of Thermoelectrics*, Rowe, D.M. (ed.). CRC Press, Boca Raton.

Kaibe, H., Kajihara, T., Fujimoto, S., Makino, K., Hachiuma, H. (2011). Recovery of plant waste heat by a thermoelectric generating system. *Komatsu Technical Report*, 57(164), 26–30.

Kuroki, T., Kabeya, K., Makino, K., Kajihara, T., Kaibe, H., Hachiuma, H., Matsuno, H., Fujibayashi, A. (2014). Thermoelectric generation using waste heat in steel works. *Journal of Electronic Materials*, 43(6), 2405–2410.

Kuroki, T., Murai, R., Makino, K., Kouji, N., Kajihara, T., Kaibe, H., Hachiuma, H., Matsuno, H. (2015). Research and development for thermoelectric generation technology using waste heat from steelmaking process. *Journal of Electronic Materials*, 44(6), 2151–2156.

McNaughton, A.G. (1995). Commercially available generators. In *CRC Handbook of Thermoelectrics*, Rowe, D.M. (ed.). CRC Press, Boca Raton.

Murase, T., Goto, D., Shibata, I., Murata, T. (2021). Thermoelectric energy harvesting vibration monitoring device KSGD-SV. *Komatsu Technical Report*, 66(173), 22–29.

Urbanitzky, A.R. (1886). *Electricity in the Service of Man*. Cassell & Company, Ltd., London.

List of Authors

Hiroyuki AKINAGA
National Institute of Advanced Industrial Science and Technology (AIST)
Device Technology Research Institute
Tsukuba
Japan

Gustavo ARDILA
Grenoble Alpes University
and
IMEP – LaHC
France

Slavko BERNIK
Jožef Stefan Institute
Department for Nanostructured Materials
Ljubljana
Slovenia

David BERTHEBAUD
CNRS-Saint Gobain-NIMS
IRL 3629
Laboratory for Innovative Key Materials and Structures (LINK)
National Institute for Materials Science (NIMS)
Tsukuba
Japan

Olga CABALLERO-CALERO
IMN-CNM, CSIC (CEI UAM+CSIC)
Instituto de Micro y Nanotecnología
Isaac Newton
Tres Cantos
Madrid
Spain

Harsh CHANDRA
University of Tokyo
Japan

Ramzy DAOU
ENSICAEN, UNICAEN, CNRS
CRISMAT
Normandie Université
Caen
France

Fazhu DING
Key Laboratory of Applied Superconductivity
Institute of Electrical Engineering
Chinese Academy of Sciences
Beijing
China

Chenguang FU
State Key Laboratory of
Silicon Materials
School of Materials Science
and Engineering
Zhejiang University
Hangzhou
China

Franck GASCOIN
ENSICAEN, UNICAEN, CNRS
CRISMAT
Normandie Université
Caen
France

Hirokuni HACHIUMA
Komatsu Ltd.
Tokyo
Japan

Sylvie HÉBERT
ENSICAEN, UNICAEN, CNRS,
CRISMAT
Normandie Université
Caen
France

Chaoliang HU
State Key Laboratory of
Silicon Materials
School of Materials Science
and Engineering
Zhejiang University
Hangzhou
China

Atsuko KOSUGA
Osaka Metropolitan University
Department of Physical Science
Graduate School of Science
Japan

Sylvain LE TONQUESSE
CNRS-Saint Gobain-NIMS
IRL 3629
Laboratory for Innovative Key
Materials and Structures (LINK)
National Institute for Materials
Science (NIMS)
Tsukuba
Japan

Airan LI
State Key Laboratory of
Silicon Materials
School of Materials Science
and Engineering
Zhejiang University
Hangzhou
China

Zhongxin LIANG
Department of Physics and Texas
Center for Superconductivity
(TcSUH)
University of Houston
USA

Antoine MAIGNAN
ENSICAEN, UNICAEN, CNRS
CRISMAT
Normandie Université
Caen
France

Marisol MARTÍN-GONZÁLEZ
IMN-CNM, CSIC (CEI UAM+CSIC)
Instituto de Micro y Nanotecnología
Isaac Newton
Tres Cantos
Madrid
Spain

Takao MORI
National Institute for Materials Science (NIMS)
WPI-MANA
and
Graduate School of Pure and Applied Sciences
University of Tsukuba
Japan

Satoru NAKATSUJI
Department of Physics
University of Tokyo
Japan
and
Institute for Solid State Physics
University of Tokyo
Japan
and
CREST
Japan Science and Technology Agency (JST)
Saitama
Japan
and
Trans-scale Quantum Science Institute
University of Tokyo
Japan
and
Institute for Quantum Matter and Department of Physics and Astronomy
Johns Hopkins University
Baltimore
USA

Mathieu PASTUREL
Institut des Sciences Chimiques de Rennes
UMR CNRS 6226
Université de Rennes
France

Zhifeng REN
Department of Physics and Texas Center for Superconductivity (TcSUH)
University of Houston
USA

Akito SAKAI
Department of Physics
University of Tokyo
Japan

Hongjing SHANG
Department of Physics and Texas Center for Superconductivity (TcSUH)
University of Houston
USA
and
Key Laboratory of Applied Superconductivity
Institute of Electrical Engineering
Chinese Academy of Sciences
Beijing
China

Junichiro SHIOMI
University of Tokyo
Japan

Bin XU
University of Tokyo
Japan

Congcong XU
Department of Physics and Texas Center for Superconductivity (TcSUH)
University of Houston
USA

Qi ZHANG
State Key Laboratory of
Silicon Materials
School of Materials Science
and Engineering
Zhejiang University
Hangzhou
China

Tiejun ZHU
State Key Laboratory of
Silicon Materials
School of Materials Science
and Engineering
Zhejiang University
Hangzhou
China

Index

3D-AAO, 87, 88

A, B

additive manufacturing, 189, 214, 223, 224
aging, 186
anodic alumina, 78
anomalous Nernst effect, 117–120, 124, 126
Berry curvature, 120–123, 125

C, D

coating, 6, 8, 9, 13
coefficient of thermal expansion (CTE), 9, 11, 14, 15, 28–31, 36, 39
contact(s), 194, 197, 225, 226
  resistance, 21, 22, 24, 27–29, 32–34, 36–38, 40
conversion efficiency, 176, 178–182, 185, 186, 189, 190, 192, 194, 199
device failure, 4, 16, 21, 24, 29
doping, 99, 110

E, F

electronic properties, 54, 55, 57–60, 64, 67
energy harvesting (EH) (*see also* thermoelectric (TE)), 242–244, 248
figure of merit (ZT), 180–186, 188, 189, 193, 197–199

G, H

geometrical parameters, 191
grain boundary, 53–55, 57–59, 62–64, 67
heat flux sensors, 124, 125, 127
hybrid harvesting devices, 175

I, J

intercalation, 100–103
interface design, 26, 28, 34
interfacial thermal conductance, 143, 157
intrinsic defects, 54, 60, 62
joining technology, 16, 30

L, M

low thermal conductivity, 132, 152–155, 160–163
low-cost thermoelectric materials, 183, 199, 213
magnetic Weyl semimetals, 121
magnetoresistance, 104
magnetothermopower, 104, 105, 109
measurements at the nanoscale, 82
mechanical properties, 186
metamaterials, 89
microdevices, 82, 86
microstructure, 53–55, 67

modulated structure, 215, 217
modules, 180, 185, 186, 189, 191–199, 214, 223, 225

N, P

nodal lines, 120, 123, 125
phonon
  anharmonicity, 154
  coherence, 161, 163
  dispersion, 135–137, 143, 150–152, 154, 159
  harmonicity, 159
  mean free path (MFP), 152, 156, 163
  scattering, 134, 139, 142, 149, 150, 152–155, 157, 161, 163
point defects, 53, 54, 58–60, 62, 67
processing, 187, 188, 213, 214
pyrite, 101

R, S

radioisotope thermoelectric generators (RTGs), 176, 181, 190, 195, 198
scattering mechanism, 55–59, 63, 64
silicides, 183, 189, 196, 213–215, 218, 219, 223–226
single parabolic band, 54, 55
solid solutions, 55, 59, 60, 67
stacking faults, 217
sulfides, 99, 101, 104, 105, 110
synthesis methods, 189

T, U, W

templates, 78, 80–82, 84, 86, 87
texturation, 103, 110
texture, 53, 54, 58, 59, 65–67
thermal
  conductivity, 73, 74, 76, 77, 80, 82, 84, 87, 89
    measurement, 144, 149, 160, 162
  cycles, 4, 5, 9–11, 22, 25, 26, 29, 33, 34, 39
  expansion, 186, 193
  spraying, 29–31
  stability, 4, 5, 8, 11, 13, 24, 40
  stress, 4, 12–15, 17–21, 32, 33
thermoelectric (TE), 53, 54, 56–60, 62–67, 131, 132, 144, 152–157, 160–165
  -module configurations, 16
  compatibility factor, 185
  conversion, 117
  cooling (TEC), 178, 181, 189, 231, 232, 234–240, 247
  devices
    structure, 190
  energy harvesting, 184, 189
  generators (TEGs), 86, 87, 89, 176–181, 186, 190, 195–198, 241–247
  materials, 180–189, 192, 195, 197–199
  technologies, 175–177, 179–182, 184, 195, 198, 199
    applications of, 176, 180, 181, 195, 198
topological magnets, 118–120, 126, 127
transition metal chalcogenides, 99
upscaling, 223
waste heat recovery, 241, 242, 245–247
Weyl cones, 120, 122

Summary of Volume 1

Preface
Hiroyuki AKINAGA, Atsuko KOSUGA and Takao MORI

Part 1. Introduction to Materials Development

Chapter 1. Strategies for Development of High Performance Thermoelectric Materials
Takao MORI, Atsuko KOSUGA and Hiroyuki AKINAGA

1.1. Introduction
1.2. Selectively lowering the thermal conductivity
1.2.1. Utilizing nanostructuring and defects
1.2.2. Utilizing crystal structure and bonding
1.3. Enhancing the Seebeck coefficient/power factor
1.4. Outlook for materials development
1.5. References

Chapter 2. Computational and Data-Driven Development of Thermoelectric Materials
Prashun GORAI and Michael TORIYAMA

2.1. General theory
2.1.1. Boltzmann transport theory
2.1.2. Relaxation time approximation
2.1.3. Thermoelectric properties
2.1.4. Defect theory
2.2. Applications
2.2.1. Transport calculations
2.2.2. Defect and doping calculations
2.2.3. Thermoelectric material search with high-throughput computations and machine learning

2.3. Outlook
2.4. References

Part 2. Thermoelectric Materials

Chapter 3. Thermoelectric Copper and Silver Chalcogenides
Holger KLEINKE

3.1. Introduction
3.2. Binary copper and silver chalcogenides
3.3. Ternary and higher copper and silver chalcogenides
3.3.1. Minerals based on copper and silver chalcogenides
3.3.2. Tl-containing copper and silver chalcogenides
3.3.3. Ba-containing copper and silver chalcogenides
3.4. Conclusion
3.5. Acknowledgments
3.6. References

Chapter 4. Sulfide Thermoelectrics: Materials and Modules
Michihiro OHTA, Priyanka JOOD and Kazuki IMASATO

4.1. Introduction
4.2. Materials
4.2.1. Rare-earth sulfides
4.2.2. Layered sulfides
4.2.3. Pb–Bi–S-based systems
4.2.4. Cu and Ag sulfide-based superionic conductors
4.2.5. Tetrahedrites and colusites
4.2.6. Chevrel-phase sulfides
4.2.7. Chalcopyrite
4.3. Modules
4.3.1. Colusites
4.3.2. Cu and Ag sulfide-based superionic conductors
4.4. Summary and prospects
4.5. References

Chapter 5. A Concise Review of Strongly Correlated Oxides
Ichiro TERASAKI

5.1. Introduction to electron correlation
5.2. Electronic states of transition-metal oxides
5.3. 3D transition-metal oxides
5.3.1. Co oxides
5.3.2. Cu oxides
5.3.3. Other 3D transition-metal oxides

5.4. 4D transition-metal oxides
5.4.1. Rh oxides
5.4.2. Ru oxides
5.5. Concluding remarks
5.6. References

Chapter 6. Nanocarbon Materials as Thermoelectric Generators
Tsuyohiko FUJIGAYA and Yoshiyuki NONOGUCHI

6.1. Introduction
6.2. Carbon nanotubes
6.3. Transport to materials studies
6.4. Chemical doping
6.5. Thermoelectric generators using CNT
6.6. TEG based on CNT sheet
6.7. TEG fabrication based on CNT-based ink
6.8. CNT yarn and their fabric
6.9. Conclusion
6.10. References

Part 3. Metrology of Thermal Properties

Chapter 7. Precise Measurement of the Absolute Seebeck Coefficient from the Thomson Effect
Yasutaka AMAGAI

7.1. Introduction
7.2. Absolute scale of thermoelectricity
7.3. Measurement methods of the Thomson effect
7.3.1. Conventional method
7.3.2. New measurement methods: AC–DC method
7.4. Summary and outlook
7.5. References

Chapter 8. Thermal Diffusivity Measurement of Thin Films by Ultrafast Laser Flash Method
Tetsuya BABA, Takahiro BABA and Takao MORI

8.1. Introduction
8.2. Laser flash method and ultrafast laser flash method
8.2.1. Laser flash method
8.2.2. Ultrafast laser flash method
8.3. Basic equation for data analysis
8.3.1. Response function method
8.3.2. Uniform single layer

8.3.3. Quadruple matrix
8.3.4. Thin film/substrate model
8.3.5. Temperature response after periodic pulse heating
8.4. Analysis of observed temperature response
8.4.1. Picosecond pulsed light heating
8.4.2. Nanosecond pulsed light heating
8.5. Metrological standard and traceability for measurements of thin film thermophysical properties
8.6. Application of measurement from industrial to basic physics
8.7. References

Printed and bound by CPI Group (UK) Ltd, Croydon, CR0 4YY
13/09/2026
14961068-0002